Ralf Jungclaus

Modeling of Dynamic Object Systems

Ralf Jungclaus

Modeling of Dynamic Object Systems

A Logic-based Approach

With a Foreword by H.-D. Ehrich

Accepted by the Faculty of Natural Science
of the Technische Universität Carolo-Wilhelmina zu Braunschweig
as Dissertation

Printed on acid-free paper

ISBN 978-3-528-05386-4   ISBN 978-3-663-14018-4 (eBook)
DOI 10.1007/978-3-663-14018-4

To Petra and Jan

Foreword

This book is concerned with conceptual modeling, design and specification of information systems.

Conventional information systems design starts with separating data from operations, designing each with its own collection of concepts, methods, tools – and people: conceptual modeling for the imformation structure, and program design for the application programs. The separation carries through until the final implementation: data are collected in databases and managed with database management systems, and application programs are implemented with programming languages.

This approach tends to suffer from a problem known as *impedance mismatch*. The basic paradigms underlying databases and programs – modeling, design, languages, and systems – do not fit easily together: there are incompatible type systems, data formats, operation modes, etc.

The object-oriented paradigm promises to overcome these problems: a system is viewed as a community of interacting objects, each incorporating data and operations. While we still have object-oriented programming languages incompatible with object-oriented database systems, ideas and approaches seem to converge towards homogeneous software systems dealing with both data and operations in a uniform way.

Viewing a system as a community of interacting objects does not solve all problems. Beyond the object concept, abstraction and structuring principles are needed, together with languages and methods to work with them.

So far, work in this area has been mostly pragmatic and system oriented. More conceptual and theoretical work is needed. Formal semantics and proof systems are rare which allow to formally state and verify relevant system properties. We need more conceptual clarification and unification as well as more theoretical underpinning.

This book provides a major contribution to the field, focussing on conceptual work with a sound theoretical basis. The book presents an object-oriented specification language for information systems, elaborating on ideas currently under discussion. The language offers a variety of structuring mechanisms for systems so that systems can be constructed from components which can be analysed locally.

The book is one of the rare texts that are at the forefront of the field, technically sound and precise, and well readable. May it find the attention it deserves.

Hans-Dieter Ehrich   Braunschweig, August 1993

Preface

The book presents an approach to formal object-oriented specification of information systems. The approach focuses on the early phases of system development where existing systems have to be described or systems to be developed have to be prescribed (*requirements specification* or *conceptual modeling*). Systems are considered to be reactive systems composed from objects that evolve concurrently in a discrete, event-driven way. An indispensable tool for modeling and analysis is a *rigorously defined language* over a set of *precisely defined concepts*. The definition of such a language (which is called TROLL) is the main contribution of this book.

Chapter 1 gives introductory remarks. The topic of this book is motivated and narrowed down. In **Chapter 2**, we show the different dimensions in the development process for information systems. The main section of this chapter deals with basic ideas characterizing the requirements engineering or conceptual modeling phase.

Part I is about the foundations underlying the language TROLL. In **Chapter 3**, fundamental notions of formal approaches to system specification and design are introduced. We cover the dimensions of functional, structural and dynamic modeling by reviewing characteristic approaches. Finally, concepts of object-oriented modeling and specification are reviewed. **Chapter 4** presents the semantic concepts underlying the TROLL language. The main section presents a temporal object specification logic which is the framework for the definition of the semantics of TROLL. Besides that, issues of data structure specification and object identification are discussed.

In **Part II** of the book, the language TROLL is defined. The language features are introduced using examples and the formal syntax as well as the semantics of language constructs are defined. **Chapter 5** introduces the base languages which are used to specify properties of objects. Furthermore, the structuring mechanisms of TROLL are presented informally. In **Chapter 6**, we describe the specification of objects and classes. **Chapter 7** presents the language features to specify different aspects of objects. We introduce the language features to describe specializations, roles, and composite objects. **Chapter 8** deals with constructs to connect separately specified objects to construct a system and to define subsystems and interfaces to subsystems.

In **Part III** of the book, we relate TROLL to other approaches and discuss our proposal. **Chapter 9** compares TROLL with a number of textual as well as graphical formalisms for object-oriented conceptual modeling. In **Chapter 10**, we summarize, discuss advantages and drawbacks and state subjects of further work.

The present book is the final version of my doctoral thesis which was submitted to the Faculty of Natural Science of Braunschweig Technical University in December 1992 and accepted in February 1993.

Acknowledgements   A thesis like this one is seldom an effort that is possible without the help of others. Thus, it is only natural at this point to thank a number of people for their support.

First of all, thanks to my supervisor Hans-Dieter Ehrich. He took me on board and gave me the opportunity to enter the IS-CORE project as soon as I started working as a researcher in his group. He supported my proposal for a topic and gave valuable advice. Thanks also to Cristina Sernadas who agreed to serve as external supervisor. Without her patience in answering questions about OSL it would have taken much longer to work with the logic.

Gunter Saake has had a tremendous impact on me and my work. I know him from the beginning of my studies and he initiated (along with Cristina Sernadas) the development of TROLL – I had the opportunity to participate in a memorable 2-day technical meeting in Lisboa where the core concepts of TROLL were defined. This was the basis of the work reported in this book. Gunter is the leader of the TROLL-subgroup in Braunschweig of which another member is Thorsten Hartmann. Thorsten made significant contributions to TROLL and put great efforts into the language report. Thanks to him for many discussions and for proofreading the thesis.

The work reported here has its roots in the IS-CORE project lead by Amilcar Sernadas. Thanks to him and all other members of IS-CORE and related enterprises (that have not been mentioned before), in particular to (in alphabetical order) Stefan Braß, Felix Costa, Olga De Troyer, José Fiadeiro, Peter Hartel, Gerhard Koschorrek, Udo Lipeck, Tom Maibaum, Robert Meersman, Georg Reichwein, Egon Verharen, and Roel Wieringa.

The colleagues in Braunschweig made working a pleasure. Along with those already mentioned I have to thank (again in alphabetic order) Stefan Conrad (for valuable comments), Grit Denker, Martin Gogolla (his questions made me think), Rudolf Herzig, Perdita Löhr-Richter, Karl Neumann (for convincing me to go for the PhD), Niko Vlachantonis and the former colleagues Gregor Engels, Uwe Hohenstein, Klaus Hülsmann, and Cristine Müller.

A number of students have contributed ideas while preparing theses or working as research assistants. Thanks to Guntram Hüske, Gotthard Korella, Jan Kusch, Judith Schulze, Stefan Scholz, Scarlet Schwiderski, Axel Stein, Rüdiger Stöcker, and Urs Thürmann.

Thanks also to the people that survived my talks and were putting up questions. Of particular value concerning my understanding of requirements engineering were discussions with Eric Dubois and Philippe Du Bois from Namur and with Jaques Hagelstein from Brussels. Special thanks also to Goetz Graefe and his group in Boulder, to Hele-Mai Haav from Tallinn, to Andreas Heuer and his group from Clausthal, to Gerti Kappel from Vienna, to Chenho Kung from Arlington, to José Meseguer and his colleagues at SRI, and to Susan Urban and Anton Karadimce from Tempe.

Last but not at all least Petra and Jan have been a source of energy and change. They have been suffering a bit from days and nights at work and I have to thank for their patience and their moral support. I dedicate the book to them.

All this would not have been possible without my parents.

Contents

1 Introduction

Information systems are evolving from simple data stores to systems that support complex tasks in complex problem ares. The problem areas are characterized to have a large number of dependencies and localities along with complex procedures. Furthermore, we observe that local information systems in organizations are being integrated in order to fulfill the need for accessing information from several sources directly. The *correctness* of data is essential in such a scenario.

Information systems in large organizations can be regarded as *reactive systems*: they provide a number of *services*, they *react* upon interactions with the environment, their behavior is influenced by the environment, and they react according to rules or procedures that are defined within the organization. Often, an information system must process complex service requests of long duration instead of short transactions with few dependencies.

The importance of modeling in the development of such information systems is increasing due to the complexity of applications and organizations. A large information system creates a virtual reality and decisions are made based on the information and support available from such systems. Thus, it is absolutely essential that such a system models *correctly* the relevant aspects that have been identified in the requirements analysis phase. It is at least as important to do the right things than to do things right. This form of correctness *cannot be verified formally*. Therefore, the development of an information system is an *interactive process* between the developers and the customers. This interactive process should be carried out over a *precise framework of modeling concepts* (cf. [TM87, Chapter 1] and [Gog92]. A proposal of such a framework is the main contribution of this thesis.

In software engineering, it has been shown that the most severe flaws are produced in the analysis and design phase [Par90, GJM91, Gog92]. Mainly, the errors are produced when transforming user needs and organizational structures and rules into a requirements specification. A lot of errors can also be produced in the subsequent phases but misunderstandings and flaws in the analysis and design phase often are not realized before the shipping of the software system to the client, which makes them very expensive to remove and increases the so-called maintenance costs. Our approach focuses on the early phases of information system development where the task is to *model* user need and organizational structures and processes. By making available a formal framework of modeling we aim at improving the quality of early specifications. It has been stated in [Gog92] that precise languages can help in the process of producing requirements specifications by reducing misunderstandings.

Modeling of systems is based on a particular *perspective* which provides interpretation concepts for a portion of the world (which implies that we are able to filter out the irrelevant aspects of the problem domain [Gri82, Lin90]). This portion of the world is called the *problem domain* or *Universe of Discourse*. Initially, the problem domain includes not only concepts to be computerized but also concepts that make up the environment of the system to be implemented (*closed systems view*). We employ a perspective that has been influenced by the conceptual modeling perspective (both structural and behavioral) [RC92] and by the original *object-oriented* perspective as proposed in the SIMULA approach [BDMN73]. The characteristics of our perspective are the following:

- The problem domain consists of a collection of *phenomena* [Lin90]. A phenomenon can exist physically or in the mental world.

- Each phenomenon is embedded in an environment. That is, a phenomenon can be regarded to be a *component* in a system. Each component has *local properties* that can be observed.

- Our perspective of the world is *discrete*. That is, we abstract from reality in that we regard state transitions to be discrete events. This is not considered harmful since a digital computing system usually is used as the implementation platform for an information system.

- The behavior of a component depends on *interactions* with its environment. Since we are working in a discrete setting, the behavior of a component can be regarded as being *event-driven*. Furthermore, an information system can be considered not to halt (at least ideally). Such systems are called *reactive systems*.

- Components conceptionally evolve *concurrently* to each other. Concurrency further increases the complexity of the system model.

In order to increase the quality of *reactive, composite information systems* that are modeled according to the perspective presented above, the modeling process should be carried out over a framework with the following characteristics [GJM91]:

- *Incremental* techniques must be possible. An incremental technique supports the stepwise development and refinement of an initial design. Incrementality is essential in requirements engineering since specifications emerge due to interactions between analysts and clients and requirements continually change [Gog92].

- A modeling framework must provide a formal semantics—only then we can ensure preciseness and in principle it is possible to prove properties of the model being constructed.

- The framework must support the *structuring* of a specification according to the perspective described above. The structuring mechanisms help in organizing large specification documents and should be usable to decrease the complexity of analyzing specifications by supporting a strong notion of locality.

- The framework must support the formal modeling of concurrency aspects. The modeling of concurrency is an active field of research in theoretical computer science and its integration with approaches to modeling structure and functionality of systems is a rather new field.

In this book we are proposing an approach along these lines. We aim at supporting *rigorous* and *object-oriented* modeling at the conceptual level. Furthermore, the modeling of static and dynamic aspects shall be integrated and concurrency shall be inherently supported. The approach shall support both the declarative and the operational style of modeling, i.e. attributes as properties observable in a state shall be syntactically and semantically distinguished from *events* or *actions* that describe transitions between states. *Locality* of properties (be they static or dynamic) is very important since it supports the structuring of specifications and changes are kept local to the smallest unit enclosing the changes without affecting the properties of other components.

Integrating components into a larger systems requires the support of bottom-up modeling. We have to provide means to connect separately specified components and we have to be able to hide unnecessary details of a component description to the outside.

An *expressive language* providing declarative and operational constructs is the vehicle we are using as core component of our approach. It is called TROLL. TROLL is much more usable than a pure formal calculus due to syntactical sugar and strong separation of specification aspects. The language provides a number of *abstraction mechanisms*:

- Specialization as a more detailed view of a concept (*is-a relationship*);

- Roles as a temporary dynamic specialization; and

- Composite objects as objects being constructed from others (*part-of relationship*).

The language also provides constructs that support putting together components of a system. The constructs also allow the extension of systems and some modularization above the object level. The constructs are:

- Relationships defining global constraints, global interactions and scripts;

- Interfaces provide restricted views on separately defined units in order to hide unnecessary details to other system components.

The semantics of TROLL is defined using a *logic*. The logic is called *Object Specification Logic* and was developed by A. Sernadas, C. Sernadas, and J. F. Costa from INESC Lisbon [SSC92]. Basically, the logic supports the specification of temporal behavior and has an inherent notion of locality. The advantages of a logical approach are

- the possibility of incremental specification, i.e. a specification can be refined or completed by adding specifications of additional properties, and

- we may derive further knowledge about properties of the system from a specification using the inference rules of the logic.

The concepts underlying the language TROLL origin in work being performed in a project sponsored by the CEC under the name IS-CORE (Information Systems— COrrectness and REusability). The basic ideas have been developed by A. and C. Sernadas along with H.-D. Ehrich in [SSE87]. There, an algebraic framework for object-oriented specification of database applications was proposed. Since then, work has been done towards more sophisticated semantic domains for object-oriented specification [ESS89, ES89, SE90, EGS90, ES91, EGS91], logical foundations based on semantical concepts [SFSE88, FSMS90, FM90, FM92, SSC92], different specification languages on the basis of the semantical frameworks [CSS89, SFSE89, SSG$^+$91, JSS91, SJ91a, SJ91b, JSH91, HJS92, Saa93] and work on methods of object-oriented specification [Ver91]. Overviews about the IS-CORE approach to object-oriented conceptual modeling give [SFSE89, SE90, SS91, SJE91,

ESS92, EDS93]. The TROLL language itself is based on previous work reported in [SFSE89, CSS89, JSS91, SJ91a, SJ91b]. A precursor version of the language has been defined in the language report [JSHS91]. This version is referred to as TROLL91 in this book. For the semantics, we used the recent paper introducing the object specification logic [SSC92].

Outline of this Book

In the next chapter, we will outline the context and the motivations for the approach described in this thesis. We will briefly discuss system development in terms of phases, quality criteria, and principles and put some emphasis on the description of principles underlying requirements specification, conceptual modeling, and object-oriented modeling.

Chapter 3 presents basic notions of approaches to system specification. Our approach borrows notions and principles from different formal approaches to specification. We will briefly introduce them and will discuss deficiencies of the approaches for the integrated specification of structure and behavior.

In Chapter 4 we will present the main semantic concepts. The chapter gives a brief introduction into the concepts of the underlying data universe. In the main part, we present the Object Specification Logic (OSL) after [SSC92]. This logic is the framework for defining the semantics of the language TROLL.

Chapters 5 to 8 make up the body of the thesis where the language is defined. In Chapter 5, we define the base languages. Sentences of the base languages are used to specify different kinds of properties in different areas of the specification structures. Furthermore, we give a light introduction to the high-level structuring concepts provided by TROLL. In Chapter 6, the specification of templates as basic structuring concept of system specifications and the specification of classes is described. Chapter 7 introduces the abstraction mechanisms specialization, roles, and composite objects after the introduction of the underlying concepts of referencing in collections and interactions. Chapter 8 finally introduces language features to put together specification units to system specifications. In each section, the language features are introduced informally before their formal syntax and their semantics is presented.

Chapter 9 compares TROLL to a number of related approaches, both textual ones and graphical ones. In Chapter 10, finally, we discuss the approach presented in this thesis and point out further research areas.

2 Information Systems Development

This chapter is about the development of dynamic, composite information systems. In the first section, we will give a brief overview about general development issues for software systems. We will sketch development phases and development models. The first section closes with a list of quality criteria in the development of software systems and principles that are considered essential for successful system development.

After this general introduction to system development we will enter into a particular phase, the *requirements specification* or *conceptual modeling* phase. We will show the tasks in this phase and will briefly compare the notions of requirements engineering and conceptual modeling. The different dimensions of techniques for high-level modeling are pointed out. We describe object modeling as a particular technique and show the impact of knowledge representation and deductive approaches on modeling and development.

The chapter closes with an introduction to our modeling approach. We will narrow down the requirements and goals of the approach presented in the remainder of the book.

Of course we will not elaborate on all aspects of system development. We are not considering organizational aspects and human aspects in software development although we know that these issues are as important as the ones sketched in this thesis. For more details see e.g. [GJM91].

2.1 The System Development Process

Usually, the development of software systems (like information systems are) is divided into a number of phases. Each phase is concerned with problems on a certain level of abstraction in the development of software systems.

2.1.1   Development Phases

The traditional development phases can be characterized as follows [GJM91, Gog92]:

- The initial stage of the development process is a definition of the *objectives*, mostly in terms of the qualities to achieve. The problem must be analyzed at least at a global level. This phase is mainly performed at an executive level. Furthermore, alternative possible solutions (and the costs in terms of money and time) are to be identified. Sometimes this phase is called the *feasibility study*.

- In the *requirements analysis and specification* phase, the qualities of the application must be defined. This should be done in a *declarative way*, i.e. the specifier must define *what* qualities the system ought to have and not *how* these qualities can be achieved. *Requirements engineering* is often defined to be the process of capturing user requirements and to *construct a model* of the system *and its environment*. This phase will be examined in more detail in the next section in this chapter. It should be mentioned that this phase is recognized to be critical in the development process of information systems because requirements are difficult to capture and they change over time.

- The *design and specification* phase should deliver a description of the system architecture (i.e. the modules, their functionality and the relationships between modules). Furthermore, the behavior of each module must be specified. The designer must also make tentative resource decisions and strategy decisions.

- In the *construction and validation* phase the system components are realized and assembled to form the system. The components are validated against the component specifications and the system is validated against the system specification. Often, this is done in a progressive way by integrating more and more components and validating the resulting subsystem [GJM91].

- In the *deployment and maintenance* phase, the system is delivered and installed. Maintenance means on the one hand to correct remaining errors that are only revealed after delivery of the system and on the other the continuous upgrading and modification of the system. Often, modifications result from misunderstandings in the requirements phase, but a system must also be modified according to changing requirements. Often, the cost of maintenance makes up to 60% of the total costs of software [GJM91].

The phases sketched above are those which can be found in a traditional system development process. When the phases are strictly separated and follow subsequently, the process is referred to as the *waterfall model* [GJM91]. It cannot be considered adequate since the requirement that the results of each phase be frozen is not realistic—there is always feedback in later phases (especially when clients use the system) that has an impact on the results of earlier phases (see Figure 2.1).

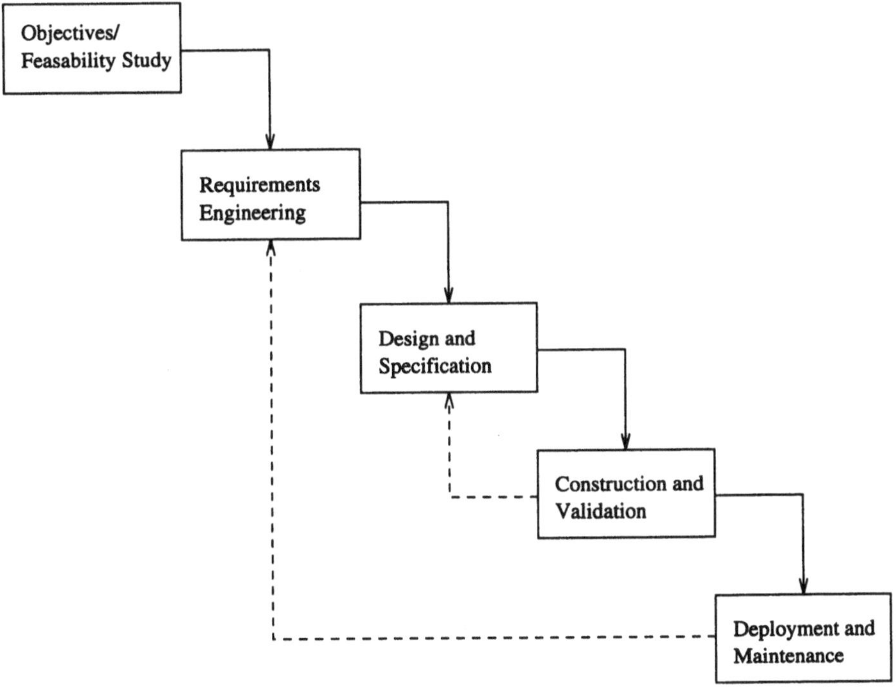

Figure 2.1: Phases in the Traditional Software Development Process

Therefore, a number of alternative process models have been proposed:

- The *incremental* or *evolutionary* model supports the delivery of *increments* to the user in order to making the user drive the evolution of the increment by operational experience. An increment is a self-contained unit along with its specifications and other supporting material [GJM91, Section 7.1.2].

- The *spiral model* consists of cycles consisting itself of four steps:

- The identification of the objectives of the software system, alternative solutions and constraints on the alternative solutions;

- the evaluation of the alternative solutions, maybe using prototypes or formal analysis;

- a plan for the next level of prototyping; and

- a review with suggestions for the next cycle.

The spiral moves outward from the center, where the costs are increasing as the radius of the spiral increases. The progress can be determined by angular displacement.

2.1.2   Quality Criteria

A number of quality criteria can be established for system development [GJM91]. The first set of criteria concerns the design:

- *Adequacy*
  An information system should represent all the desired or even needed aspects of the problem domain. That is, the requirements having been recorded in the requirements analysis phase must be correct in the sense that they are a faithful representation of what the system must ensure in order to satisfy the needs of the client organization. Obviously, there is no way of measuring this quality and it is not possible to guarantee adequacy. In our opinion, the only way to get as close as possible to what the client expects is to perform the process of capturing requirements in an *interactive* way, if possible by using *prototypes* or *animated specifications*, i.e. by simulation. We consider this criterion to be very important since many errors and flaws do not come in during design but during requirements elicitation.

- *Correctness*
  A system is correct if it does not violate its specification. This definition implies that we must be able to *formally verify* whether or not a system meets its specification. An indispensable prerequisite for doing that is the availability of a formal specification of the system or of certain aspects of the system. From these considerations it is also clear that formal representation techniques should be used as early as possible in the development process of information systems.

- *Verifiability*

  The properties of a system should be easy to verify against the specification. To achieve this, formal analysis techniques should be used since testing cannot guarantee the correctness of a system. Generally, formal techniques can hardly be used due to the complexity of the task. Therefore, the structuring of a system into units that can be verified in isolation is considered to be useful to ensure verifiability.

- *Modifiability*

  As said before already, requirements change over time. A system should be easy to modify to meet the new requirements. Today, such changes are often not started at design level but at the product level. Due to the complexity of software systems, the impacts of such changes cannot be fully determined and the system becomes even more fragile. Therefore, the original system design must be performed with modifiability in mind and all changes to the system should be performed at the level of design or even at the level of requirements specification.

- *Reusability*

  In our setting, reusability at the requirements level is most important. During the modeling of a new application we may try to identify parts that are similar to parts used in a previous application. We may then reuse parts of that application instead of developing a new specification from scratch. This, however, requires that the *semantics* of specifications is precise and that the impact of adaption and modification of reused parts can be fully determined. Reusability is not considered explicitly in this thesis, but the proposed approach seems to make a high level of reusability possible.

Finally, let us state two criteria that are essential for dynamic information systems:

- *Integrity*

  Integrity requirements state conditions that must be guaranteed in every state of the system or that must be guaranteed in a sequence of states. Integrity implies that there be not contradictions in states.

- *Safety*

  Safety mainly is considered in real-time systems as the absence of undesirable behaviors. It also is important in information systems since more and more information systems integrate behavioral aspects and these can cause severe problems (like e.g. in financial applications). Safety requirements state what should always be guaranteed by a system in execution.

Naturally, there are many more quality criteria for software systems. Since we are only considering the abstract specification of systems and problem areas, we will not mention implementation-related criteria.

2.1.3   Development Principles

There are a number of principles that can result in better systems and a successful development of systems. Principles are not enough to drive the process of system development but they are considered essential in the development of large software systems. The list is by no means exhaustive, but it covers the important aspects of information systems development on the conceptual level.

- *Rigor*

  A specification should be as formal as possible. On the one hand, formality induces preciseness. On the other hand, formality makes possible the derivation of knowledge from a design and thus makes the analysis of specifications possible. Full formality, however, is hard to use and understand, in particular by clients who must evaluate a conceptual model specification. In any case, even if a less formal notation is used, there should always be a rigorous semantics to ensure preciseness and analyzability.

- *Modularity*

  A complex system usually consists of a number of subsystems or components. On the conceptual level, there are a number of goals that require modularity: we must be able to decompose a system into components, we must be able to compose components to form a system, we must be able to understand components in isolation in order to reduce the complexity of a system, and we must be able to encapsulate components in order to separate concerns. Decomposing means to divide the system into subsystems in order to concentrate on the local properties of subsystems first. Compositionality means to be able to put subsystems together (using appropriate means to specify their interaction) and yield a semantics of the composite system. The ability to analyze components separately reduces the complexity of the analysis task and supports modifiability. Encapsulation of modules means that there must be a precisely defined interface to a components only allowing access to relevant properties (*black box*). The object-oriented perspective towards systems design supports modularity on the object level. This, however, is too fine a granularity when it comes to the design of large systems.

- *Abstraction*

  Abstraction means to identify the important aspects of a system and to ignore irrelevant details. In conceptual modeling and requirements specification, it is necessary to abstract from implementation-related details of the solution. When we develop a specification, we also abstract from reality. It is generally not possible to capture all details of a problem domain. We have to decide upon which aspects are relevant and which are not. This implies that there may be many different abstractions from the same reality (often called *views* [Per90]).

- *Anticipation of Change*

  When designing or modeling a system, the designer should be aware of the fact that there will be changes in the problem domain that have to be reflected by the model, maybe even introduced through the use of the system in an organizational environment. The principle of anticipation of change should be used to achieve modifiable specifications and systems. A technique following this principle is structuring a design into encapsulated units. Other aspects like version control and related issues are not considered in this thesis.

- *Incrementality*

  Incrementality claims that we may develop a design in a stepwise manner, where each step is called an increment. Thus, we must be able to expand an initial design by additional properties and must be able to refine a design in that more details are added. The motivation for incrementality is that requirements usually emerge while analyzing the problem domain and while interacting with the client about conceptual models [Gog92].

After having discussed properties of the design process especially in the early phases, let us now enter into a particular phase in the development process, the requirements analysis and specification phase.

2.2 Requirements Analysis and Conceptual Modeling

Among the first phases of information systems development we can find the requirements analysis or conceptual modeling phase. An important task in this phase (besides cost estimates, resource estimates, and development time estimates) is the modeling of the problem domain. A model is an abstraction of something in order to understand it before constructing it. Basically, the main tasks are:

- an analysis of an organization or a system in order to understand and describe that organization or system precisely, and

- an analysis of the interaction and relationships of the system with its existing or desired environment.

In [DHR91, DDP93], the relationship between the tasks for requirements engineering have been illustrated as shown in Figure 2.2. In the *elicitation task*, the relevant information about the problem domain is collected from the customers or users by the analyst. The techniques for elicitation are not considered in this book. The information collected in the elicitation task is processed in the *modeling task*. The result of the modeling task is a *specification of a model* in the above sense (i.e. a formal representation of the information collected in the elicitation phase) or an improved version of an already existing specification. In the *analysis task*, the specification is (formally) analyzed for consistency and (if possible) completeness. An indispensable tool in the modeling and analysis tasks is a *rigorously defined language* over a set of *precisely defined concepts*. In the *validation task*, a model is presented to the customers or users for examination. The primary task of validation is to check whether the specified model meets the requirements—this can neither be ascertained nor be proven by any modeling and analysis framework.

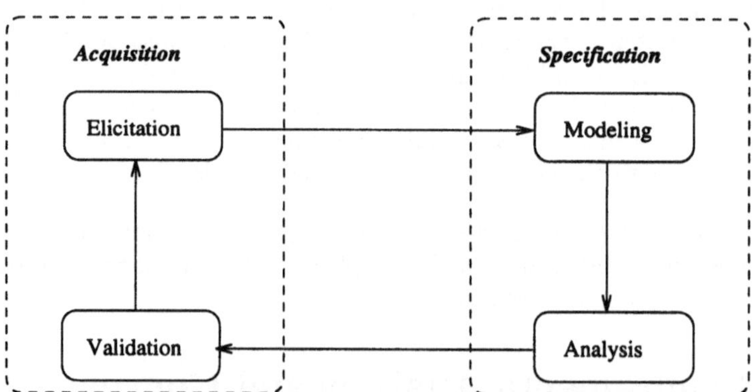

Figure 2.2: The Requirements Engineering Tasks

Requirements engineering for dynamic information systems must also analyze the interactions of components to be computerized and the environment. In the requirements document, we are likely to have references to manual procedures to be installed, to real-world entities, or to specific devices (like sensors). Such systems are often called *composite systems* [Fea87, DD92]. Only some components of a composite system represent software (sub-)systems. Other components are represented by structures to be controlled or monitored in software components. Of special importance are the associations and interactions between software components and other components of the system—they are turned into interfaces of the software system to its environment later.

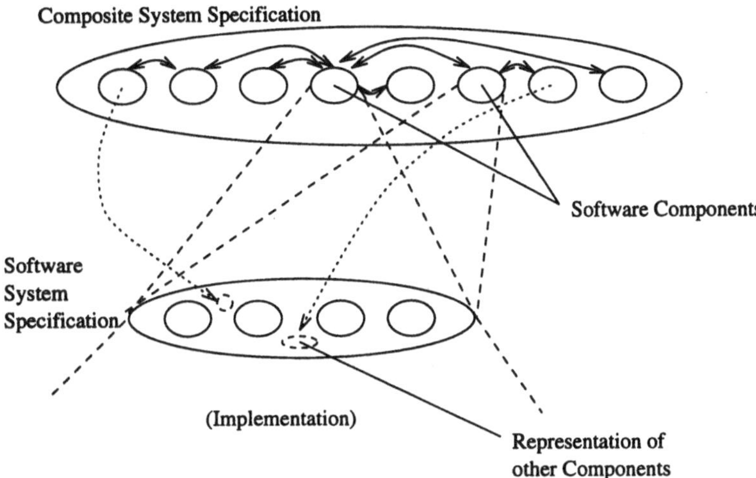

Figure 2.3: A Composite System Specification and its Implementation

Figure 2.3 illustrates the notion of composite system. It shows the composite system specification which consists of interacting objects. Some of these objects represent software systems—they are defined and refined in the specification of the software system. This specification also includes representations for those components that interact with the software system or are controlled by the software system.

2.2.1 Notions and Properties

The term *requirements analysis* usually denotes the process of capturing the requirements of the client. The end-product of this phase is a *requirements specification* which contains as its most important part a specification of a model of the problem domain. Traditionally, this model is *functional*, i.e. it identifies the functionality which is to be computerized by the system to be developed. The data structures are developed around the functions.

The term conceptual modeling denotes a similar process in the context of information systems. A conceptual model is usually *data-oriented*, i.e. it identifies the data to be represented and its structuring according to *entities* in the problem domain. The functionality of the system based on the data structures usually is vaguely defined. The end-product of conceptual modeling usually is called a conceptual model (although it really is a *specification* of such a model).

Recently, it has been realized that the temporal aspects of an information system (in particular the evolution of data over time) were not treated sufficiently by neither the functional nor the data-oriented approach. The *behavioral perspective* (cf. [Lin90, Saa91, DHR91, RC92]) therefore focuses on the dynamic evolution of data and the impact of events on the state of the system.

In any case, a specification describes a model which enables us to understand the problem. It contains the relevant aspects and properties of a portion of the problem domain.

Qualities

We want to concentrate here on the qualities that a specification must have. Many of the qualities have been mentioned already for the entire development process. We briefly want to comment on the most important ones in this phase:

- *Modifiability*

  Since the requirements specification is the vehicle for understanding a problem, it is also the basis for interactions between developers and clients. It has also been said before that requirements emerge gradually in the analysis phase. Thus, a specification is to be considered as an increment which is subject to refinement.

- *Understandability*

  A specification must be understandable. For large specifications, it is essential that we are able to analyze parts of the specification in isolation. Ideally, the specification parts correspond to components of the system. In order to understand the whole specification, we must be able to compose the parts into a specification of the whole system.

  Another important aspect of understandability is the *expressiveness* of specification formalisms. A developer should not be forced to code already at the conceptual level but should be able to express things as natural as possible. A specification in a formal calculus or in terms of mathematical expressions is hard to understand for the developer and impossible to understand for the client. The use of graphical symbols also helps a lot.

2.2.2 Techniques

A great number of techniques have been proposed for requirements specification. In the following, we want to mention some techniques and evaluate them against some issues.

Structural and Behavioral Modeling

In a specification of a model of the problem domain, we must be able to specify *structural and behavioral aspects*.

Behavioral modeling mainly is used in connection with Structured Analysis techniques [DeM79]. Structured Analysis techniques analyze a problem domain by discovering the data flows in the problem domain. Data is transformed and transformations are connected. The basic concepts of *Data Flow Diagrams* are

- *external entities* which act as sources or sinks of data outside the system to be computerized;

- *data transformations* which represent *functions* mapping input data to output data;

- *data stores* that store data, and

- *data flows* between external entities, data stores, and transformations.

The emphasis lies on the modeling of the data transformations, hence such approaches are called "functional". The process of producing a Data Flow Diagram is called functional decomposition, since it decomposes a problem domain according to the data transformations being identified or desired. Data is grouped around functions. The approach *neglects the structure* of the problem domain. Furthermore, functions are more likely to change than entities [GJM91].

Other behavioral modeling approaches emphasize *states* and *state transitions*. A prominent example are *Statecharts* [Har88].

There have been proposed a large number of formal approaches to behavior specification. On the one hand, we have several approaches in the *process specification* corner, of which the basic ones are CCS [Mil80] and CSP [Hoa85]. A process is considered to be a collection of possible life cycles. On the other hand, we have Petri nets [Rei85], which allow for the specification of concurrent behavior.

In general, approaches to behavioral modeling allow for at most very restricted ways to model structural properties. The formal approaches neglect the need for structural modeling except for LOTOS [ISO84] which employs algebraic specification of data types.

Structural modeling in contrast concentrates on the identification of the entities in a problem domain, their properties and the relationships between entities. These are the main concepts of the Entity-Relationship model [Che76] of which a number of extensions have been proposed, e.g. [HNSE87, EN89, Hoh90]. Structural modeling is also underlying other database models.

In structural modeling, the description of operations on structures or of behavior is usually not integrated. These aspects have to be described in a different model. In general, coherence of models is a problem when using such an approach.

Thus, none of the approaches is suited to cover both structural and behavioral aspects satisfactorily. One solution would be to *combine* the specification of different aspects of a problem domain. It has been shown, however, that neither starting with functional decomposition and take data stores and external entities as input for a structural modeling approach nor starting with a structural approach and defining behavior using data flow diagrams leads to satisfactorily results [SF91, Wie91b, Wie91a].

A promising approach is the object-oriented approach. There are approaches, however, which use different diagrams for the structural, behavioral, and functional aspects of a system which fail to integrate fully these aspects later [RBP+91].

Declarative and Operational Specifications

Behavior can be specified in several styles [HR92].

- The *declarative* style employs formulas to *constrain* the possible behaviors. Thus, the goal is to specify the valid states, the valid changes, and the *admitted behavior*. The approach provides a very high level of abstraction along with the possibility to describe locality of properties. The severe drawback is known as the *frame problem* [HR92]: we do not only have to specify the required changes, we also have to *explicitly exclude the undesired ones*! If we specify that certain conditions imply the change of the value of a variable, then we might want that *only* under these conditions a change is allowed; this must be specified explicitly by *excluding all other possible changes*.

- The *operational* style employs explicit control through calls and triggers to execute operations that modify information. There is an implicit assumption that no change takes place unless induced by the execution of an operation. This approach, however, prevents an abstract description of admissible behavior and constraints. It is e.g. impossible to express a requirement like "salaries never decrease" directly; we have to analyze all operations to be able to infer such a requirement.

In the requirements engineering phase for dynamic information systems we are likely to encounter various kinds of behavior. For some kinds, the declarative style has advantages over the operational style, for others the operational style is more suited. Thus, it has been argued in [SE90, FSMS90, JSHS91, SSC92] as well as in [HR92] that reconciling both approaches results in a powerful approach to specifying the dynamics of information systems.

Top-Down and Bottom-Up Modeling

The usual way in requirements analysis is to employ a *top-down* approach [DD92]. One starts with identifying *global goals* which the system as a whole must fulfill. Then, the components of the system are identified and to each of them the local responsibilities and properties are attached such that the decomposition preserves the original global specification. Each component can be further decomposed by applying the previous step recursively.

Bottom-up modeling means to analyze the problem in order to identify required components, to select suitable components from a library and adapt them or create them according to the needs, and to assemble the components to form the system.

In practice, the goal is to combine both approaches: in order to understand a sufficiently large problem domain, one has to decompose it into subsystems which are then analyzed in isolation. The relationships between the components must be analyzed, too. Then components are modeled and assembled using the relationships identified before. Such a process is supported by *object modeling.*

2.2.3 Object Modeling

The notion of object-orientation was coined with the presentation of the language SIMULA [BDMN73], which was originally developed for simulation. The developers of SIMULA took as a basis for their language the idea for regarding the situation to be simulated as a *collection of objects that interact.* This idea has been employed by many approaches to modeling, among them OBLOG [SFSE89] or OMT [RBP+91].

A central characteristic of problem domains is their *concurrency.* In has been pointed out in [Weg90], that object-based concurrency allows the natural modeling of concurrency aspects in problem domains and thus expands the power of modeling approaches. In a more formal setting, even concurrency alone is an active field of research. Combining concurrency and object-orientation *in a formal setting* is still an active field of research. First attempts are e.g. [Agh86, SEC90, EGS90, ES91].

Concepts

Usually, the following characteristics are agreed upon in research on object-oriented concepts:

- *Object*

  An object is the *basic structuring unit*. It represents a concept with sharp boundaries. It is described by an interface consisting of observable properties (attributes) and operations or events as state changes, and admissible behaviors. An object description describes actual *instances*. An instance is in a *state* at each moment of existence. The state is reflected in the values of the attributes.

- *Identity*

  Each instance has its own unique identity. Thus, we are able to distinguish instances in a state even if the values of all their observable properties are identical. We must also be able to refer to the same object in different states.

- *Classification*

  Objects having the same properties and showing the same admissible behavior can be grouped into a class. A similar notion is that of *typing* although the direction is the opposite: a type describes the potential instances. The currently existing instances form a class of that particular type.

  The main purpose for classification is abstraction—we do not have to describe each instance separately, it deserves to describe a generic instance and the possible extensions of classes.

  Please note that the classification of instances is not necessarily based on the syntactical notion of having the same properties and behavior—classification is always performed according to problem-related semantics.

- *Specialization*

  Objects (and classes) can be related to others to form a hierarchy. A specialization instance is a more detailed view on the same conceptual entity than the base instance. Therefore, it is called an *aspect* in [ES91]. On the specification level, the specified properties of the base also hold for the specialization but may be further refined. On the instance level, the base instance is included in the specialization instance. Please note that this ensures that both instances refer to the same conceptual entity.

Specialization is a static concept: if it exists, a specialization exists for the whole lifetime of the base aspect, and there can exist at most one specialization of a kind of a base instance.

- *Roles*

  Role instances are *temporary specializations*. In contrast to specialization, there can be multiple roles of one base instance in a state and a role can be played several times in the life of a base instance. Roles are a new concept that origins in the modeling of behavior-intensive applications [Sci89].

- *Aggregation*

  Composite instances or *aggregations* have other instances as components, they are *parts*. The properties specified for the parts or components remain valid in the context of the composite object. An access to components, however, is only possible through their interface—direct manipulation is not possible.

- *Relationships*

  When considering objects as the basic building blocks of system models, we have to be able to specify interactions and dependencies between them. Following the tradition of semantic data modeling, this should be separated from the description of the objects itself. It is not desirable to describe context-sensitive information (as relationships to other components are) in the local context of an object description. Relationships have been proposed as a modeling concept in object modeling in [RBP+91].

Implications

Object modeling on the conceptual level implies the following deviations from traditional modeling principles [RBP+91, Boo90, Ver91]:

- The emphasis is not on the modeling of the functionality of the entire system or the structure of the application domain but on the structure and behavior of the system components, the objects. That is, information is localized around objects.

- There is no separation of structural and operational specifications. Objects are highly independent entities containing highly coherent (structural and behavioral) information local to it.

- The behavior of the system is determined by the behavior of the components and the interaction between them. There is no global control instance describing an invocation structure for functions.

2.2.4　Knowledge Representation

Knowledge representation is concerned with *world modeling* [Myl90]. "Objects" are regarded as the things that populate the problem domain or Universe of Discourse.

Researchers interested in knowledge representation are interested in *symbolic structures* that represent information. These structures (also called *notation*) must be suitable to be used to describe (rather than prescribe) open, hardly definable and unpredictable domains.

Very early, researchers have found predicate logic to describe assumptions holding in problem domains very useful. Predicate logic, however, results in a clumsy description since it is difficult to organize descriptions. A good overview on the use and limitations of logic in knowledge representation can be found in [Kow89]. Logic is only a formalism to be used at some internal level because it does per se not provide any structuring concepts geared towards external (i.e. real-world) semantics.

Today, one can identify three main streams of notational approaches to knowledge representation [Myl90]:

- *Entity-based* notations
  Such notions offer some kind of entity concept. An entity is assumed to have a one-to-one correspondence to an "object" in the real world.

- *Term-based* notations
  These notions focus on the specification and organization of terms used to express knowledge about a problem domain. Term-based notations are often equipped with description-forming operations. Terms are interpreted in a symbolic way in the tradition of deductive logic.

- *Frame-based* notations
  Frames are data structures to represent common knowledge. They furthermore support different forms of common sense reasoning, including default reasoning.

A particular representative of entity-based notations are semantic nets [RM75]. A semantic net is a directed graph whose nodes denote entities or entity types and whose labelled edges denote relationships of several kinds:

- `is-instance-of`
  The target node of an `is-instance-of` relationship is an entity type of which the source node is an instance. This implies that the description of the schema and the actual population are mixed in one semantic net.

- `is-a`
  An `is-a` relationship denotes *specialization*: the source entity is a specialization of the target entity.

- `is-part-of`
  This relationship describes the fact that the source entity is regarded to be a component of the target object (which is then an aggregation).

Some semantic data models like TAXIS [MBW80] are based on semantic nets. Frame-based notations [Min75] employ the following concepts:

- Frames are data structures denoting a *natural kind*, i.e. they are similar to an object type. A frame has a number of *properties* which apply to the kind of type and a number of *slots* which are attributes or procedures denoting a data value. Slots are local to an instance of a frame.

- There is an *is-a* relationship between frames along with *inheritance* of properties and slots from the base frame. We can have monotonic inheritance as well as non-monotonic inheritance (in the presence of exceptions and defaults).

- Integrity constraints, like e.g. cardinality constraints, can be attached to frames and relationships.

Frames focus on concepts of stereotypes and default reasoning. The world is modeled by comparing newly discovered "objects" to already known stereotypes. In absence of information, default assumptions are made which can be retracted in a non-monotonic way when more information becomes available.

Concepts from entity-based and frame-based knowledge representation approaches have influenced requirements specification and data models. In particular the inference techniques that come with formal approaches to knowledge representation have had a significant impact on data models. We only want to mention the notions here:

- Closed-World Assumption (CWA)
  A representation of a problem domain always leaves certain aspects unspecified. The closed-world assumption basically states that unspecified facts have to be considered as being false. This way, an incomplete representation can be completed. For details see [Bra92].

- Unique Name Assumption (UNA)
  An unambiguous representation of knowledge must rely on the assumption that names for entities be unique.

- Non-Monotonic Reasoning
  Non-monotonic reasoning basically means that assumptions that have been made in the absence of facts can be dismissed when information is available (which is not true in classical logic).

Moreover, the representation of *temporal* and *spatial* knowledge has become very important when regarding domains with inherent dynamics. There is a variety of approaches that penetrates many fields from planning [AHT90] to information and data modeling. Recent developments in the context of information systems modeling are RML [GBM86] and Telos [MBJK92].

These considerations conclude the discussion of the context of research in system development.

2.3　Requirements and Characteristics of a Logic-based Approach

We have pointed out the importance of a rigorously defined language for modeling and analysis in the early phases of information systems development. The concepts underlying that language shall

- support the perspective of object systems, i.e. a system is represented as a collection of interacting objects that evolve over time in a discrete, event-driven way, and

- support the abstract specification of systems according to that perspective.

We are trying to combine characteristics of the approaches to the development of complex systems described above in order to come up with a new language that fulfills the following requirements:

- The language shall support a *problem-oriented* way of modeling. That is, we want to model problems and their conceptual solution, but not the implementation of the solution.

- This implies that the language must have a high degree of *expressiveness* because the specifier should not be forced to encode the problem to be able to specify it.

- The language must have a well-defined semantics, which should be as rigorous as possible. The semantic framework must be *logic-based* in order to make reasoning about specifications possible.

- The language must support both a declarative and an operational style of specification in order to be used for capturing *structural and behavioral* aspects of systems. Thus, we must integrate the description of states and evolution of states as well as the description of events or actions that cause states to change.

- The language must provide a wide variety of structuring mechanisms for specifications in order to reduce the complexity of a specification. The structuring mechanisms should make use of *object-oriented* concepts. The structuring mechanisms should not only support better understanding of system specifications but should also make possible to reason about components in isolation.

Part I
Foundations

3 Basic Notions of System Specification and Modeling Approaches

In this chapter, we want to give a brief introduction into the basic concepts of different approaches to system modeling. We cover a wide variety of approaches ranging from algebraic specification over semantic data modeling and modeling and specification of reactive and concurrent systems to the basic notions in object-oriented modeling and specification. Concepts from these approaches will be used in the approach presented in this thesis.

3.1 Algebraic Specification

Algebraic Specification is a specification technique that supports the abstract definition of *abstract data types* (ADTs). An ADT describes a class of data domains and operations on these domains by listing the services available on these data structures along with the properties of these services [VL89, EMO92]. The goal is to abstract from the implementation or concrete representation of data structures and operations. Algebraic specification provides an important basis for the approach described in this thesis.

The specification of ADTs thus means the definition of classes of *algebras* that model the ADT. A specification consists of two things:

- A *specification Spec* of common properties and invariants that have to be fulfilled by an algebra, and

- a *semantic construction* used to define the class of algebras that fulfill a given specification.

We will put emphasis on the concepts for specification and will just give a brief summary of semantic constructions.

An algebraic specification $Spec = (\Sigma, E)$ consists of a *signature* Σ and a set of *equations E*.

- A signature $\Sigma = (S, \Omega)$ itself consists of a set of *sort symbols S* and a set of *operation symbols* Ω. A signature introduces names for the sets of elements in an algebra (sorts), and names for single elements (*constants*) and functions of an algebra. An operation is defined over sorts $s_1...s_n, s$ in the form

$$op\colon s_1, ..., s_n \to s$$

where $n \in I\!N_0$. Constants are defined as nullary operations, i.e. they represent constant functions that denote a single value in a set.

- An *interpretation structure* over a signature is an algebra $A(\Sigma) = (A(S), A(\Omega))$. $A(S)$ associates with each sort $s \in S$ a *carrier set* $A(s)$, $A(\Omega)$ with each constant $c\colon \to s$ a value $v \in A(s)$ and with each other operation $op\colon s_1...s_n \to s$ a function $f\colon A(s_1) \times ... \times A(s_n) \to A(s)$.

- *Terms t* are constructed from sorted variables and operation symbols. A term has an associated sort and denotes a value of that sort.

- *Equations* $t = t'$ are defined between terms t, t'. Equations can be combined to form *conditional equations* of the form $t_1 = t'_1 \wedge ... \wedge t_n = t'_n \Rightarrow t = t'$.

Using equational reasoning and induction, we may infer theorems from the equations (see e.g. [VL89]).

As an example consider the (very simple) specification of the data type of integer numbers. In some ad-hoc notation, this would be specified as follows:

```
data type integer
  sorts integer;
  opns
    zero:→ integer;
    succ:integer → integer;
    pred:integer → integer;
    add:integer × integer → integer;
  eqns
    vars i,j:integer;
    pred(succ(i)) == i;
    succ(pred(i)) == i;
    add(zero,i) == i;
    add(succ(i),j) == succ(add(i,j));
    add(pred(i),j) == pred(add(i,j));
end data type integer;
```

Most specifications assume *comparison* operators to be present. Therefore, they include a sort bool containing the truth values **true** and **false**. A complete specification for the data type BOOL would be

```
data type BOOL
  sorts bool;
  opns
    true,false:   → bool;
    not:  bool → bool;
    and,or:  bool × bool → bool;
  eqns
    vars a,b,c:bool;
    a and a == a;
    a or a == a;
    a and b == b and a; a or b == b or a;
    a and (a or b) == a;
    a or (a and b) == a;
    a and true == a; a or false == a;
    a and not a == false;
    a or not a == true;
    not true == false; not not a == a;
    (a and b) and c == a and (b and c);
    (a or b) or c == a or (b or c);
    a and (b or c) == (a and b) or (a and c);
    a or (b and c) == (a or b) and (a or c);
  end data type bool;
```

Most approaches use *many-sorted algebras*, i.e. algebras contain a family of sets of data elements. A variant are *order-sorted algebras* [FGJM85, GKK⁺87] where a partial order is defined on the sorts reflecting the fact that one set of data elements may be included in another, e.g. the sort **nat** of natural numbers is contained in the sort **integer** of integer. Operations then can be restricted to subsorts and still are the same operations. Such sorts are referred to as *subsorts*.

Basically, research has focused on two ways to construct the semantics of specifications:

- *Loose semantics* associates with a specification the class of all algebras that satisfy the equations in E.

- *Initial semantics* associates with a specification a class of algebras that are equal up to isomorphism. The algebras thus cannot be distinguished by properties expressible in equational logic. Initial algebras do only contain elements that are denoted by terms over the signature (*no junk*) and terms denoting the same data element can be proven to be equal using equational reasoning (*no confusion*).

Pure initial semantics is often considered too restrictive, whereas loose semantics suffers from little expressiveness, i.e. too many interpretations are accepted. Therefore, proposals have been made for constructions between these extremes (cf. [Wir90]).

Algebraic specifications of larger systems can be *structured*. The mechanisms allow to construct larger specifications from smaller units. Usually, the following mechanisms are considered:

- *Extension* allows to add sorts, operation symbols, and equations to a given specification.

- *Union* constructs the union of given specifications with shared sub-specifications. An example is the construction of a specification `List` by putting together a specification of `BasicList` (including only the most general operations for generating an empty list, appending elements to a list, and concatenating lists), a specification for natural numbers, and a specification for boolean values `bool` and extending the resulting specification by an operation denoting the length of the list.

- *Renaming* allows us to have bijective renaming of sorts and operation symbols in specifications.

All these mechanisms are based on the notion of *signature morphism*. A signature morphism between signatures Σ_1 and Σ_2 is a pair $\zeta = (\zeta_S, \zeta_\Omega)$ of functions between sort symbols and operation symbols, respectively, such that the following condition holds:

$$\text{if } op : s_1, ..., s_n \to s \ \in \ \Omega_1 \text{ then } \zeta_\Omega(op) : \zeta_S(s_1)...\zeta_S(s_n) \to \zeta_S(s) \in \Omega_2$$

A further structuring mechanism is *parameterization*. Parameterization allows to specify *schemes*, i.e. specifications that can be instantiated by *parameter specifications*. This allows us to specify data structures that share common principles, e.g. sets, lists, or tuples of which the concrete specification only depends on the sort of elements. The instantiation is described by a signature morphism that defines the replacement of formal parameters by actual parameters.

For a more detailed presentation of the field of algebraic specification see e.g. [EM85, VL89, EGL89, Wir90].

In general, algebraic specification supports the structuring of data. To a certain extent, it can be used to specify requirements, but due to the pure mathematical approach it is more suited for the (functional) specification of programs.

A severe drawback of the algebraic approach is that the concept of state is not supported. Thus, it is not directly suited to model objects and object systems. Only by auxiliary constructions it is possible to model objects that persist and evolve over time. Nevertheless, the algebraic approach has been used successfully as a basis for object-oriented specification approaches, among them OBLOG [SSE87] and FOOPS [GM87].

3.2 Semantic and Object-Oriented Data Modeling

In the late seventies it was recognized that the modeling power of traditional data models was not sufficient enough for high-level modeling of database applications and information systems [Ken78]. Data models support a system-oriented view of the data and neglect how the sources of data are structured in the problem domain. Data modelers were required to encode those structures in terms of the few modeling constructs the data model provided. Consequently, the result was a poor database design, which could only be maintained if a good documentation was available. There were some approaches to extend traditional data models with more constructs. The most popular approach was the RM/T Model by Codd [Cod79] which was based on the relational data model. Therefore, researchers came up with a number of proposals for improved data models (which were called *Semantic Data Models* later).

Semantic Data Models

One of the first proposals (and up to now one of the most popular ones) was the *Entity-Relationship Model* proposed by Chen [Che76]. Other proposals were TAXIS [MBW80], SDM [HM81] and the IFO-Model [AH87].

It has been recognized early that in spite of the popularity of the ER-Model the modeling constructs were not rich enough for conceptual modeling. Therefore, a number of so-called *Extended ER Models* have been proposed. Their additional modeling constructs are often based on the suggestions in the article [SS77] and are called *data abstractions* in the context of semantic data models.

A common principle of almost all semantic data models is the support of semantic hierarchies. That means that these models support a number of abstractions mechanisms. Abstraction mechanisms are special relationships which have a meaning in problem domains.

- *Classification* means the clustering of objects with the same properties into classes. The properties define the *type*. Often, a relationship *is-instance-of* is defined between an object and the type.

- *Aggregation* means the composition of objects from component objects. Aggregation is defined both on the level of objects and on the level of types: on the type level, it is the product of the component types.

- *Grouping* means to group collections of objects and represent a group as a single entity. Objects of a group type can be regarded as elements of the powerset over the base type.

- *Specialization* and *Generalization* [SS77] allows to express subset or *is-a* relationships between entities.

It was only recently, however, that an approach towards a formal algebraic semantics has been proposed along with an extended ER model [HG88, Hoh90, GH91, Hoh93]. The ideas that lead to the development of this model are reported in [HNSE87, EGH+92] and are one of the basic sources on which many concepts of our approach rely.

The main concepts of that EER-Model are the following [EGH+92]:

- *Entity Types E*:
  An entity type describes possible collections of objects with common properties. Entities describe abstract, non-printable objects in a Universe of Discourse. Only the properties of entities are printable and are represented by the values of their attributes.

- *Attributes a*:
  Each attribute belongs to an entity type and represents a property of an entity of the type. Attributes are typed, i.e. their values range over a data sort. There are several special kinds of attributes in the EER-Model:

  - *Data valued* attributes range over data sorts, which may be user-defined.

  - *Object-valued* attributes range over entity types, i.e. their possible values are entities of the associated type.

  - *Multi-valued* attributes range over parameterized types, i.e. over lists or sets of values/entities.

Attributes by default are optional, i.e. they may have an undefined value.

An attribute a is represented as a function $a: E \rightarrow s_a$, where E is the associated entity type and s_a is the sort of the attribute.

In the EER-Model, attributes allow to model complex structured entities. Component objects can be described through object-valued attributes, grouping is expressible through multi-valued attributes.

- *Relationship Types $r(E_1, ..., E_n)$*
  Relationships are the usual form of aggregation in ER-approaches. A relationship type defines possible relationships between entities. A relationship instance r relates entities $e_1, ..., e_n$ of types $E_1, ..., E_n$, respectively. A *rolename* is a label for a link associating a relationship with a participating entity type.

- *Type Constructors*
  A type constructor can be used to define new classifications for entities of particular entity types. No inheritance is defined by a type construction. By means of derived attributes the attribute values of input entities can be carried over to entities of of constructed types.

  Type constructors are used to express various types of specialization/generalization relationships:

  - Simple *specializations* (i.e. subtypes) are single types constructed from single input types. Each entity of a specialization type is also an entity of the base type (but not vice versa).

  - A *partition* is the construction of several *disjoint* specializations of a single input type with an additional constraint that for each entity in the base type there must be an entity in one of the constructed types.

  - *Generalizations* are single types constructed over several input types.

An example of an EER-Specification is the specification of a banking world depicted in Figure 3.1. Both **ATMs** and **Accounts** are object-valued attributes of the **Bank** type, i.e. a **Bank** instance has as components a set of **ATM** instances and a set of **Account** instances. There is a partition of **Account** into **SavingsAcct** and **CheckingsAcct** such that each **Account** instance is an instance of either **CheckingsAcct** or **SavingsAcct**. A subset of the **Person** instances may be specialized as **Customer** instances.

Semantic Data Models originally were designed with the purpose of capturing concepts of a proposed information system, i.e. they were tailored towards the

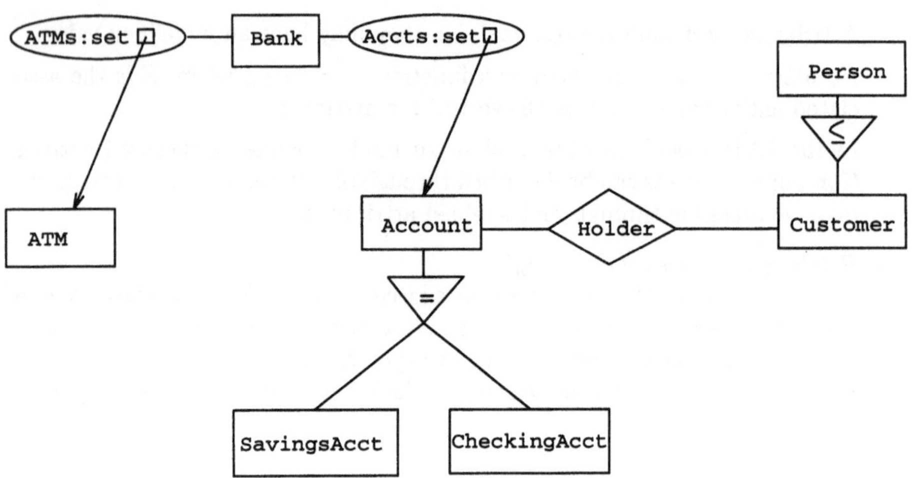

Figure 3.1: An Example for an EER-Specification

design of a (software) system. Only the proposed extensions or abstraction mechanisms as sketched above make them suitable also for the modeling of composite systems.

As pointed out in [GBM86], traditional Semantic Data Models are only capable to define *admissible snapshots* of a problem domain. Thus, those approaches do not address any dynamic aspects of the problem domain. This, however, is essential when it comes to modeling at the requirements level.

In some Semantic Data Models attempts have been made to integrate the modeling of dynamic aspects. In TAXIS, the notion of *transaction* is integrated in the model [MBW80]. A transaction can be regarded as being a function defined by preconditions, actions, and postconditions. If a transaction cannot be executed to fulfill pre- and postconditions, an exception handling mechanism is invoked.

The following example (from [Won81]) shows a transaction specification for a seat reservation on a flight.

```
transaction RESERVE-SEAT with
     params reserve-seat:(p,f,d);
     locals
       x:   RESERVATION;
       p:   PERSON;
       f:   FLIGHT;
       d:   DATE;
```

```
prereqs
   seats-left?:  f.seats-left > 0;
actions
   make-reservation:
       insert-object x in RESERVATION with
           person:p, flt#:f.flight#, date:d;
   decrement-seat:  f.seats-left <- f.seats-left-1;
end
```

After the declaration of parameters and local variables, the preconditions for the execution of a transaction are specified in the **prereqs** section. In the **actions** section, the actions to be executed inside the transaction are executed in the order of specification. Optionally, postconditions may be specified in an **results** section. In the example, however, no postcondition is stated.

In the approach described in [Saa91, EGH$^+$92], the structural model is combined with temporal integrity rules describing temporal dependencies and with actions. The result is a four-level approach:

- On the *data type level*, data values and operations on them are specified using an algebraic approach.

- On the *object level*, an extended ER-Model is used to specify the states of a model in terms of entities and relationships between entities.

- On the *object dynamics level*, two means are provided to specify the evolution of entities:

  - In the *evolution component*, the evolution of properties over time is described in a declarative way. As an example consider the following temporal integrity constraint over the EER-Schema of Figure 3.1:

    ```
    var (a:Account)
    (a.Balance < 0) implies (sometime a.Balance >= 0)
    ```

    It states that an `Account` in the red (i.e. with a negative `Balance`) must have a positive `Balance` again sometime in the future.

  - In the *action component*, admissible actions changing the state of the model are specified in terms of pre- and postconditions or in a procedural way.

    As an example for a declarative specification of a transaction consider the following one describing the deletion of an `Account`:

```
transaction DelAccount(Acct#:natural)
on          exists (a in Account) a.No = Acct#
pre         a.Balance = 0
post        not exists (a in Account) a.No = Acct#;
```

The transaction can only be applied if the enabling condition in the **on** section holds in the current state of the database. It has no effect if it can be applied but the precondition under **pre** does not hold in the current state; otherwise, the effect is stated as a postcondition under the keyword **post**.

- On the *process level*, actions can be combined. This level has not been worked out in detail.

Good overviews of the large number of semantic data models give the articles [HK87] and [PM88]. The dynamic aspects in connection with semantic data models are discussed in [Bor85, UD86]. One of the major drawbacks of most proposals were the poor (if not totally missing) formal semantics. Only few approaches have a formal semantics, among them TAXIS [Won81], and the EER-Model [Hoh90, GH91, Hoh93].

Object-Oriented Data Models

A plethora of object-oriented data models has been proposed for a number of object-oriented database systems which were developed in the last five years [ZM89, Heu92]. It has been argued, however, that (in contrast to the situation in value-based data models) a generally accepted object-oriented data model cannot be presented. We want to briefly cite the basic requirements for object-oriented data models as proposed in [ABD+89] and [Bee90].

Object-oriented data models strive to integrate the concepts of object-oriented programming and semantic data modeling. This is not an easy task since there is one fundamental difference between object-oriented programming approaches and semantic data models [Kin89]:

- Object-oriented programming models support *behavioral abstraction* whereas

- semantic data models strive to provide *structural abstractions*.

Other differences lie in the aggregation mechanism. In object-oriented programming and data models, aggregation is often simulated by references whereas aggregation over static objects is a first-class concept in semantic data models.

Some object-oriented data models are mixing the notions of structured value and object. An object can be described by a structured value, but the object identifier clearly distinguishes objects from values [Bee90]. Many object-oriented data models assume object identifiers to be generated and assigned by the system. This is not adequate for a data model in which the modeling concepts are implementation independent [Bee90].

The basic notions in object-oriented data models are the following:

- *Type constructors* for the construction of complex values for the structured description of objects;

- *Object Identity* (to be discussed in Section 3.5.2) for the representation of objects;

- *classes* and *types* for classification—both concepts should be separated. A type describes the structure of possible instances and a domain of possible instances. Classes denote sets of instances of which the extension may vary over time. The extension of a class is always a subset of the possible extension defined by the associated class.

- *Inheritance* can be regarded both on the type level and on the instance level. In contrast to object-oriented programming models, most object-oriented data models support the inheritance of attribute values, i.e. the inheritance of instances.

- *Relationships* between classes are often just simulated by object references, i.e. by pointers to objects in terms of attributes having object identifiers as values. There is a number of data models, however, that support composite objects (e.g. ORION [KBC$^+$87]).

- *Integrity constraints* are often neglected in object-oriented data models. Usually, only key constraints or cardinality constraints are supported, but not arbitrary integrity relationships.

A major drawback of most object-oriented data models is the concentration on the description of static structures. Updates and state transitions are neglected in the semantics of object-oriented data models.

Object-oriented data models usually provide a number of generic operations that have been motivated by database languages. Generic operations can be applied to objects regardless of their type.

In contrast to object-oriented programming models, object-oriented data models support a weaker notion of encapsulation. As known from data models, attributes can be accessed directly.

A central issue in object-oriented data models are declarative database languages. On the one hand, a database must provide a *query language*, on the other hand the *impedance mismatch* [ZM89] between today's database languages and programming languages for the development of applications on top of a database must be avoided.

Object Logics

Based on the concepts of object-oriented data models and *deductive database languages*, a number of *object logics* have been proposed (see the annotated bibliography [CG92]).

Object logics basically are logical calculi where the concepts of object identity, type or class hierarchies, and inheritance are integrated. Object logics concentrate on the deduction of information (virtual objects, derived attributes, queries) form the *current state* of a collection of objects. Thus, the emphasis lies on the *grouping of information around objects*. State properties (represented e.g. by attributes) are not encapsulated in the sense of object-oriented programming where only methods can be employed to refer to the local state.

Object logics cover mainly static aspects. Only recently approaches to integrate *updates* have been proposed (e.g. [Abi90, KLS92]). Even these approaches, however, fail to offer capabilities for the description of temporal behavior and evolution.

With these consideration we leave the basic concepts of *structural* modeling and turn to approaches for the *behavioral* modeling, which are the other extreme.

3.3   Models of Reactive and Concurrent Systems

Let us now turn to approaches that neglect the structural description of a problem domain and concentrate on the description of the dynamic behavior. Of particular interest are approaches that support the modeling of non-sequential behavior.

A number of models for non-sequential systems have been proposed in the last years. In this section we will briefly review important basic concepts that help in understanding the dynamic aspects of our approach.

There are several ways to model non-sequential systems. The conceptual differences can be characterized as follows:

- Interleaving vs. concurrency
  Interleaving means to represent parallel processes by an interleaved occurrence of the atomic actions of both processes in one sequence. That is,

concurrency is reduced to nondeterminism. In general, a given concurrent run of several processes in parallel can be represented by many different possible interleaved orders of atomic actions. There is an ongoing debate in the theory of concurrency whether representing concurrency by interleaving is adequate and realistic. In fact, there are distinct execution sequences of concurrent systems that do not result in different interleaved representations. Compared to concurrent models, models based on interleaving have a simpler theory of specification and verification which makes them easier to use for such approaches.

- State-based vs. event-based representations
  Models can consider states or state transitions (or better *events* causing state transitions) as their primary modeling primitives. In general, state transitions (or actions or events) are considered to take no time for execution, they are atomic.

- Global vs. compositional models
  There are different perspectives on systems to be modeled: we can model the *global* behavior of an entire system or we can concentrate on the *local* behavior of system *components*. Then the overall behavior of the system results from the composition of the local behaviors. Almost all traditional models like transition systems, event structures, and Petri nets are global. Models based on message passing are going one step in the direction of local behavior specification as do models that allow for the simultaneous execution of state transitions in a global model based on interleaving. Recently, models based on object-orientation have gained some significance like the ACTOR model [Agh86].

Transition Systems

A transition system is a global model which is often considered as a general model of concurrency [Mil90, MP92a]. The basic elements of a transition system are the following:

- A finite set of *state variables* $v_1, ..., v_n$ which are defined to range over domains.

- A set of *states*. Each state is an interpretation of the state variables, assigning to each variable a value over its domain.

- A finite set of *transitions*. Each transition τ is mapping a state s into a (possibly empty) set of states $\tau(s)$ that can be obtained by applying τ to s.

Each state $s' \in \tau(s)$ is called a τ-successor state of s. The set of transitions must include an idling transition mapping a state s to itself for every state.

- An initial condition characterizes the initial states in the transition system, i.e. those states at which a process may start.

We furthermore have a *transition relation* for each transition τ which relates the values of the state variables in a state s to the values of the state variables in a τ-successor state s'. The transition relation consists of

- an *enabling condition* stating the condition under which the state s may have a τ-successor and

- a conjunction of *postconditions* defining the values of state variables in s', possibly in terms of their values in state s.

A *run* or computation of a transition system is a sequence of states $s_0\, s_1\, s_2 \ldots$ such that

s_0 is an initial state

each state s_{i+1} is a τ-successor of state s_i for some transition τ.

A run can be represented as

$$s_0 \xrightarrow{\tau_0} s_1 \xrightarrow{\tau_1} s_2 \longrightarrow \ldots$$

A state s is *reachable* in the transition system if there exists a run containing s.

Transition systems can be represented by *transition diagrams* (see [MP92a, Section 1.2]). More details on transition systems can be found in [MP92a] and in [Mil90] where labelled transition systems are presented along with rules defining their properties.

Message Passing

A more abstract model is that of message passing. In this model, processes communicate through channels. Message passing implies that there be *synchronization* between processes because a message cannot be received before it has been sent.

We will restrict ourselves to models using synchronous communication only which underlie e.g. CSP by Hoare [Hoa85] or CCS by Milner [Mil80, Mil89] and neglect problems of local time as discussed in [LL90].

In general, there is a set C of typed channel names or port names. The set of actions is extended by the following special types of actions:

- A *send action a!t* sends the value denoted by a term t via channel a; send actions are sometimes called output actions.

- A *receive action a?x* assigns a value on channel a to the variable x of the appropriate type.

In the presence of synchronous communication, channels are not capable of buffering the values send to them. Thus, when processes communicate via a channel, they are *synchronized* through a corresponding pair of send and receive actions that are permitted to occur simultaneously.

Synchronous communication is a useful abstraction from the actual realization using acknowledgements and other things. On the other hand, though, the approach does have problems in describing the actual behavior of distributed systems. If the process which is specified to receive a value from a channel is not able to do that the sending process cannot progress anymore. Such problems are discussed in [LL90].

Problems implied by Interleaving

When representing concurrency by interleaving, some problems arise. We want to mention these problems very briefly.

The most severe problem is that in a interleaved run the independent progress of each concurrent process is not guaranteed. This is due to the need to choose *one* of the enabled transitions of any of the processes. Thus, we may have the situation that processes may be starving, i.e. only transitions of other processes are continuously chosen preventing any progress in the starving processes.

One solution for this problem is to impose further restrictions on interleaving models to ensure progress for every process participating in an interleaved run. Such restrictions are usually known under the name *fairness* requirements [Fra86].

Fairness problems do not arise in concurrent models which will be discussed in the sequel of this section.

Other problems arise when events or actions are refined. Some possible interleaved runs must be forbidden if they cause effects which violate the more abstract specification.

Petri Nets

Petri nets are models of concurrent systems. There are many variants of Petri nets [Rei85]. We will present the basic concepts of place-transition nets (P/T-nets), an abstraction of simple Petri nets. A P/T-net consists of a set P of *places*, a set T

of *transitions* and a relation $F \subseteq (P \times T) \cup (T \times P)$ which is the set of edges. A P/T-net is usually represented by a bipartite directed graph where the edges connect places (represented as circles) with transitions (represented as squares) and transitions with places. The pre-set of a transition is defined as the set of places connected to the transition by edges. The post-set of a transition is the set of places to which a transition t is connected. Another important concept is a *marking* $\vec{M}$. A marking is a vector associating with each place $p \in P$ a natural number. Graphically, a marking is represented by tokens drawn into places—for each place p_i with marking $\vec{m}[i]$ we draw $\vec{m}[i]$ tokens inside the circle representing p_i.

The basic operation on marked nets is the *firing* of a transition. In order to fire, a transition must be enabled. A transition t is enabled under the marking $\vec{M}$ if each place p_i in the pre-set of t has a marking $\vec{m}[i] \geq 1$, i.e. iff each place in the pre-set of t has at least one token. The firing of a transition t removes a token from each place in the pre-set of t and adds a token to each place in the post-set of t. In general, there is no restriction on the order of firings. Each enabled transition may fire.

Petri nets have been applied in the behavior specification for software systems. They are, however, a global model. Moreover, the tokens do not contain any information and thus the model is purely control-oriented.

Event Structures

Event structures basically model causality between occurrences of events. An event is an action which is regarded to be indivisible. There are many variants of events structures. We will just give the basic notions here. More details can be found in [Win89].

An elementary event structure consists of a partial order $(E, \leq)$ where E is a set of events and $\leq$ denotes causal dependency: if $e \leq e'$ then e' depends on the previous occurrence of e. A state of computation in an elementary event structure is the set of events that have occurred so far in the computation. This set is left-closed and finite, i.e. $\forall e \in E$: $\{e' \mid e' \leq e\}$ is finite denoting that there be only a finite number of events causing e to occur.

Elementary event structures do not consider nondeterminism which might occur when modeling interaction with the environment. An improved model are prime event structures, where a binary conflict relation $\# \subseteq E \times E$ is added modeling that the occurrence of certain events rules out the occurrence of others. A configuration then is a subset $l \in E$ which is left-closed and does not contain conflicting events.

Although the model is rather intuitive, some authors have shown that the model cannot be used when system behavior is to be synthesized from the behavior specifications of the components [PP92].

Concurrent Rewriting

A recent proposal by Meseguer presented a logic called *concurrent rewriting logic* [Mes90, Mes92]. Basically, this logic consists of

- a *Rewrite Theory* R

- and a set of deduction rules.

A rewrite theory R consists of

- a *signature* (Σ, E) where Σ is a family of function symbols indexed by its arity and E is a set of Σ-equations;

- a set L of *labels* and

- a set of *rewrite rules* R with elements $r: [t] \to [t']$ where $r \in L$ and $[t], [t']$ are elements of E-equivalence classes of terms over Σ and a set of variables X.

A rule $[t] \to [t']$ should be read as "$[t]$ becomes $[t']$". A term should be understood to be a proposition that asserts being in a certain *state* having a certain structure. The rules of a rewrite theory thus can be used to reason about change in a concurrent system.

Concurrent rewriting, i.e. the actual operational view, coincides with deduction using the following rules:

1. Reflexivity
   For each term class $[t]$ holds $\vdash [t] \to [t]$.

2. Congruence
   For each function symbol $f \in \Sigma_n$

   $$[t_1] \to [t'_1]...[t_n] \to [t'_n] \vdash [f(t_1, ..., t_n)] \to [f(t'_1, ..., t'_n)]$$

3. Replacement
   For each rewrite rule $r: [t(x_1, ..., x_n)] \to [t'(x_1, ..., x_n)]$

   $$[w_1] \to [w'_1]...[w_n] \to [w'_n] \vdash [t(\vec{w}/\vec{x})] \to [t'(\vec{w'}/\vec{x})]$$

4. Transitivity

$$[t_1] \to [t_2] \wedge [t_2] \to [t_3] \vdash [t_1] \to [t_3]$$

When a sequent $[t] \to [t']$ can be obtained from a rewrite theory R by a finite application of the above rules, it is called a concurrent rewrite.

By including associativity and commutativity in the set of equations E the logic is capable to describe concurrent rewriting of a multiset of terms, thus describing the concurrent evolution of states in a concurrent system. Details about the logic can be found in [Mes90, Mes92, Mes93].

Actors

In the Actors model (which originated in [Hew77] and has been presented in detail in [Agh86]), a system is represented by a set of *actors*. Actor systems represent a community of actors that change as the system evolves. Actors communicate through asynchronous message passing.

An actor is defined as a two-tuple consisting of

- a *mail address* with an associated *mail queue* for incoming messages and

- a current behavior which is a function of the communication being taken from the mailqueue and processed.

The behavior of an actor consists of

1. a finite set of *messages* sent to specific actors whose mail addresses are known by the actor;

2. a finite set of *new actors* created (of which the mail address initially is known by the creator only); and

3. a *replacement behavior* which will accept the next communication.

The replacement behavior differs from a state transition in that a completely new behavior is created instead of just an update of state variables. Concurrency is introduced by the ability to send several messages in response to a single message. Actors provide a strong notion of encapsulation similar to object-oriented languages. For a detailed introduction into the actor model see [Agh86].

3.4 Modal and Temporal Logics

Temporal logic usually is viewed to origin in *modal logic* which is concerned with the notions of *necessity* and *possibility*. A widely accepted semantics for modal logic are so-called *Kripke Structures*.

Kripke Structures

The basic idea behind Kripke structures is that we have a set of *possible worlds* which are related by a *reachability relation*. Formally, a Kripke structure consists of

1. a set $\mathcal{W}$ of possible worlds,

2. a reachability relation $R \subseteq \mathcal{W} \times \mathcal{W}$,

3. an interpretation function I from $\mathcal{W}$ into a suitable codomain (see below), and

4. a *current world* $w_0 \in \mathcal{W}$.

The interpretation of a possible world may be given in different ways:

- it may be an algebra $A(\Sigma)$ over a signature Σ which serves as a model for a state, or

- it may be a set of valid formulas—then the theory generated from the formulas defines the properties of the world.

The first approach is considered e.g. in [Saa88] whereas the latter has been considered in [Eme90, Section 3].

Kripke structures can be represented by directed graphs where the nodes represent worlds and the edges represent the reachability relation. Such a structure where the interpretation of worlds is not considered is sometimes called a *Kripke frame*.

Traditionally, such structures served as interpretation structures for modal logics. The modal operators $\Box$ and $\Diamond$ have been put on top of traditional logic with the following meanings:

- A formula $\Box\varphi$ states that φ must hold in all worlds that are reachable from the current world.

- A formula $\Diamond\varphi$ states that there is a world w' reachable from the current world such that φ holds in w'.

The reachability relation can also be interpreted as a temporal relation: if a state w' is reachable from a state w, then this can be interpreted as w *precedes* w'. This was a starting point for the development of temporal logics.

Temporal logics are present in many different versions in the literature. The differences lie on the one hand in the basic sets of operators and resulting syntactical differences and on the other hand on the semantic level where there are several classes of models over which temporal formulas are interpreted. In the following paragraphs we want to present basic notions and characteristics of some of these aspects.

Linear Temporal Logic

In linear temporal logic, formulas are interpreted over *sets of runs* which are sequences of states generated from some model. Thus, time is considered

- to be discrete,

- to have an initial moment, and

- to be infinite into the future.

An linear-time structure is a special Kripke structure where each world (or state) has at most one successor world and there is only one state that does not have a predecessor state: the initial state.

The basic operators (as presented in [Eme90, MP92a]) are

- $\bigcirc(\varphi)$ (the *next-operator*) stating that φ holds in the next state;

- $\square(\varphi)$ (the *always-operator*) stating that in every future state φ holds;

- $\diamond(\varphi)$ (the *eventually-operator*) stating that there is a future state in which φ holds;

- $\varphi\,\mathcal{U}\,\psi$ (the *until-operator*) stating that there is a future state in which ψ holds and in each state prior to that φ holds;

- $\ominus(\varphi)$ (the *previous-operator*) stating that there is a predecessor state and φ did hold in that state;

- $\diamondsuit(\varphi)$ (the *once-operator*) stating that φ holds in the current state or in some preceding state;

- $\boxminus(\varphi)$ (the *has-always-been-operator*) stating that φ holds in the current state and in all preceding states;

- $\varphi \mathcal{S} \psi$ (the *since-operator*) stating that φ holds in the current state or in some preceding state and ψ holds in all states following the state in which φ holds to the current state.

A light introduction into first-order temporal logic as it is used for specification purposes can be found in [MP92a].

Branching Temporal Logics

In contrast to linear temporal logic, *branching-temporal logic* considers tree-like structures over which formulas are interpreted. The motivation for such an approach is that the model may directly represent different *possible futures* thus being closer to traditional modal logic. The operators are in some respects different to the ones having been presented for linear temporal logic above.

Basically, this difference origins from the fact that we have to be able to state formulas to be valid in certain *paths* in a tree structure. A path can then be considered as being a linear sequence.

A restricted form extends linear temporal logic by *path quantifiers*. The formulas have the form

$$\mathsf{A}\varphi$$

or

$$\mathsf{E}\varphi$$

where φ is a formula of linear temporal logic with just one temporal operator. A means "for all possible futures" and E means "for some possible future". Intuitively, the path quantifier selects paths to which φ is applied.

Full branching-time logic [Eme90, Section 4] extends the limited form by allowing formulas following a path quantifier to be arbitrary linear temporal logic formulas and to allow such formulas to be connected by boolean connectives and linear temporal quantifiers.

For a detailed presentation of branching-time temporal logic please refer to [Eme90].

Past and Future Operators

In most approaches using temporal logics for the specification and reasoning about (concurrent) systems only future tense operators are used. This is reasonable when we are able to define an initial state. It can be shown that the inclusion of past tense operators does not add expressive power [Eme90]. Past tense operators are useful, though, since they make the specification of certain properties more natural and convenient.

Partial Order Temporal Logics

Even further goes the approach to consider *partial order structures* as interpretation structures. Thus, we do not consider an interleaving of state transitions but consider only precedence relationships between state transitions.

In [Rei91], directed arc-labeled acyclic graphs are considered as Kripke frames over which temporal logic formulas are interpreted. The graphs are considered to be *occurrence nets*, i.e. they refer to the actual occurrence of transitions over time. An occurrence net represents *one possible run* of some system.

An arc in an occurrence net connects two subsequent event occurrences (nodes) and is labeled by a proposition or a state. An arc represents a local state, a node in a graph represents synchronized events. Arcs imply a partial order on event occurrences, and the transitive closure of the partial order relation represents *causal dependency*: if a node e_2 (an event occurrence) can be reached on some path in the graph from a node e_1, then the event occurrence e_1 causes the occurrence e_2, and e_2 can only occur if e_1 has occurred before.

Temporal formulas are interpreted using the notion of *cut*. A cut is a *maximal set of unrelated arcs*, i.e. a collection of local states that do not causally depend on each other. Cuts correspond to virtual global states, but we do not have explicit global states. Partial-order temporal logics support distributed states.

We do not want to discuss this model further since we will work with linear time models later. Details about partial-order temporal logics can be found in [Rei91].

These considerations conclude the discussion of approaches to the specification of non-sequential systems from a behavioral point of view.

3.5   Object-Oriented Modeling and Specification

Let us now present notions of object-oriented modeling and specification. When it comes to modeling structure and dynamics of problem domains, we have to combine the approaches to modeling the behavior of non-sequential systems with the approaches to model the structure and the state changes of instances in a system. There are different approaches to achieve this—many of them emerged from the field of algebraic specification and model-based specification.

In this section, we will enter into particular issues only. Before we consider the problem zone of object identification and referencing, we will give a brief introduction to notions of object oriented specification. Finally, object specification concepts which are the basis for our approach are presented.

3.5.1 Object-Oriented Specification

Most object-oriented specification approaches are considering object-orientation on the language level only—the models are the traditional ones. Exceptions are e.g. the FOOPS approach [GM87], CMSL [Wie90], and the work underlying the present approach [SSE87, EGS90, ES91, ESS92]. In these approaches, objects are considered also in the semantic domain.

Basically, objects are characterized by the following properties:

1. Objects encapsulate a local state (that may change over time) and a local behavior.

2. Objects are distinguishable by a unique identification which persists through state changes.

3. Objects can be classified.

4. Objects can be composed to form complex objects.

5. The taxonomy of objects is reflected in the model.

These properties distinguish objects from values—values are immutable, they cannot be considered to have a state.

Property 1. means that state information is local to an object and can only be referred to through an interface. State information is usually represented by attributes with changing values.

Many approaches use an algebraic framework for objects. Objects are considered to be elements of special sorts, which are usually called *classes*. Observable properties of objects are called attributes and are modeled by functions that take an identity i and a current state σ to yield a value which is the current value of the attribute of i in state σ.

State changes are described by *events* or *methods*. An event induces the transition of an object in a state to the same object in another state. Event occurrences have effects on the values of local attributes. These effects are usually described in a declarative manner by equations or postconditions similar to the equational specification of operations in algebraic specification.

Objects evolve over time by the occurrence of events or methods. This is called *behavior*. To capture the behavior of objects, objects are considered to be *state machines* or *processes*.

The concepts are present at many places in this thesis and shall not be discussed in detail here. Let us now focus on the problem of object identification and referencing.

3.5.2   Object Identification and Referencing

The problem of identifying objects is a fundamental problem in the object-oriented approach [KC86]. It seems to be solved quite satisfactory for programming and local databases, but there are still a lot of problems to be solved when considering specifications that abstract from implementation. Further complication is raised by the various structuring concepts in specification approaches. A discussion of object identification in the context of a similar approach to conceptual model specification is discussed in [WJ91]. Another approach towards referencing objects is described in [Ken91]. After a brief list of requirements on the identity concept from our point of view we will discuss other approaches and will finally present our solution.

In the sequel, we will try to avoid speaking of object identifiers and refer to them just by using the term *identifier* in order to prevent the reader from getting confused.

Let us first list requirements we would like to be fulfilled by a concept for object identification in specification approaches:

- An identifier should allow us to distinguish different objects that are in the same state, i.e. that cannot be distinguished by regarding their state only.

- For each object we must have a unique identifier. This uniqueness should be universal, i.e. uniqueness must be valid even when system boundaries are crossed.

- An identifier must be stable, i.e. it must be possible to refer to different states of the *same* object.

- Identifiers should be *visible*, i.e. they should support accesses and references to an instance using the identifier.

- Identifiers must not carry any information that may be subject to change when instances evolve.

- It is desirable that identifiers can be assigned to newly-created instances by their environment or by a user. We think that both ways of "baptizing" objects occur in problem domains.

We are aware of the fact that it is generally not possible to fulfill every requirement listed above. In particular, it is impossible to come up with a solution for universal uniqueness of identifiers.

Let us now briefly review how the identity problem is dealt with in other approaches.

- In the data modeling field, it has been argued as early as in [SS77] that updates must preserve the integrity of individual entities. It has been seen quite early that the notion of user-defined keys (as used in the relational model) is not sufficient due to the following problems:

  - The values of key attributes cannot be allowed to change over time since then we cannot refer to different states of the same conceptual object;

  - the requirement that key values of distinct objects be distinct may lead to the introduction of artificial key attributes to fulfill the key constraints.

  In [Ken78, Chapter 3], a number of examples is mentioned which show the difficulties that arise when descriptive data is used for identification. The thoughts about a solution go in the direction of the approach given in [Ken91] which is described below.

- Surrogates are used e.g. in RM/T [Cod79] for identifying tuples uniquely within the system. After [Ken78], surrogates need not be exposed to users, their format must not be specified by users, they are atomic, they do not carry any information, and there is a one-to-one correspondence with the things they are representing. A notable operation (besides the check on equality) is the *coalesce operator* defined in [Cod79]: it allows the merging of separate objects that are realized to be the same.

- Algebraic approaches such as CMSL [Wie91c] or FOOPS [GM87] use *names* as identifiers, i.e. elements of an identifier sort. In FOOPS, names are elements of *classes* which are special sorts.

- CMSL like e.g. Glider [DDRW91] and TROLL91 [JSHS91] allow names or identifiers to be *structured values*. Thus, an identifier sort can be declared to be isomorphic to a (complex) data sort, e.g. a tuple of values of certain sorts.

- Kent [Ken91] addresses the problem at a higher level. He distinguishes *handles* or *symbols* and *tokens* which are occurrences of symbols. This view corresponds nicely to the abstract data type view of the approaches cited above: a symbol is a term denoting elements of the concrete carrier set of the identifier sort, the tokens. Kent also discusses issues of scope of handles, operations on handles and references, and validity of symbols. A crucial point in Kent's article is the construction of tokens. Here we have again a nice similarity between Kent's and the algebraic approach: construction of tokens means evaluation of terms, thus computing the denotation of symbols.

- In [KC86], the authors present an object model supporting built-in identity, i.e. identifiers are assigned to values by the system. Identifiers are assigned to objects when they are created (i.e. each existing object has a unique identifier) and there is no identifier without an object being assigned. The object model supports various notions of equality: identity denotes the actual identity of objects, shallow equality compares the values of direct components of an object and deep equality compares also the values of nested components. It also supports a merge-operator having the same meaning as the coalesce-operator in RM/T (see above).

The following table (Table 3.1) compares the approaches described above in terms of the criteria support of global uniqueness, (temporal) stability, visibility, support of assignment of identifiers by the environment and support of structured identifiers.

	unique	stable	visible	env.-assignable	structured
DB Keys	+−	−	+	+	?
Surrogates	+?	+	−	−	−
ADT-Values	+	+	−	+−	+
Kent [Ken91]	+	+	+−	+−	?

Interpretation: +   supported
+−   possible, can be simulated
−   impossible

Table 3.1: Evaluation of Identification Concepts

3.5.3   Object Specification Concepts

Before we are going to elaborate on semantic concepts underlying our approach we will give some intuitions about the envisaged structures.

Underlying objects we have at our disposal a *data universe*, i.e. data types being specified using algebraic specification techniques as outlined in Section 3.1. The data universe includes

- carrier set for data sorts (sometimes also called domains) which provide us with possible values of particular data sorts and

- operations on sorts denoting functions on domains.

The values and operations of the data universe are used to describe properties of objects.

As explained before, we consider objects to be different from values in that values are immutable (i.e. they do not have a state that changes over time), and they can be represented directly (i.e. they identify themselves). Thus, we may not represent objects directly in a system, we may only describe objects through their properties. Let us sketch the basic principles for describing objects in our approach:

- An instance is an encapsulated unit of structure and behavior. The state of an instance may be observed through an object *interface* allowing the observation of structural properties (described by *attributes*) and the observation and execution of state transitions (described by *events*). A change in the observable properties can only be induced by the occurrence of local events (*locality principle*).

- The state of an instance determines the possible observations and the possible state transitions. The possible state transitions may depend on the history of the instance, i.e. to the state transitions that occurred up to the current state. We must be able to specify permission and prohibition of state transitions in a state.

- Attributes and events are considered to be strictly *local*, i.e. they cannot be manipulated from outside a local context. Therefore, the impact of state transitions in one instance on the state of other instances can only be carried out by relating events of instances in a larger context (interaction). Attributes may only be read in this larger context but may not be manipulated by events not being local to the smallest context of an attribute.

- Relating events of instances is also the mechanism for *communication* between instances. By exchanging data in interactions messages may be send and received.

- Events can occur in parallel only if the effects of the state transitions described by them are not conflicting, i.e. they must be deterministic.

- Instance models are generic, i.e. the framework must permit to have a large number of instances of the same kind.

- Instances can be put into a larger context without violating any of their local properties (*safe object import*). This means in particular that not only the generic structure and behavior of such instances is visible in the larger context but also the current state.

- Last but not least, we must be able to specify the referencing of instances. Therefore, a concept of unique identification of instances is needed. This concept should be compatible with a concept of referencing objects through certain observable properties since this is the way we tend to reference entities in problem domains.

Let us now briefly sketch the important features of the semantical framework underlying the TROLL language which itself is described in Chapters 5 to 8.

- Each instance is uniquely identified by a data value of which the actual representation is not known. An identifier is an element of the carrier set of an *identifier sort* which provides identifiers for instances of the same kind.

- The interface of an instance is described by the elements of a *template signature*. A template signature contains (along with the identifier sorts being used) the *attribute symbols* and the *event symbols* describing structural properties and state transitions.

- An *observation* is defined through a special predicate over attribute terms denoting the fact that an attribute can be observed as having a value. An *attribute term* is an attribute symbol instantiated with terms denoting data values.

- In each state, a set of events *occur* defining a state transition. In order to occur, an event must be *enabled*. Both the occurrence and the enabledness of events are described by predicates over *event terms*, i.e. event symbols with terms denoting data values as actual parameters.

- The state of an instance is defined as a set containing predicates describing

    - the possible observations in that state,

    - the enabled events in that state, and

    - the actually occurring events in that state.

- The *admitted behavior* of an instance over time is defined by formulas of a temporal logic which describe the evolution of local states in terms of the valid predicates in states. As explained in Section 3.4, temporal logic exactly serves the purpose of describing how the validity of formulas evolves over time.

- The underlying time structure (or Kripke frame, see 3.4) is linear, i.e. states are considered to be totally ordered.

- The mechanisms to relate instances are the following:

  - *Specialization* describes a more detailed aspect of a conceptual entity. Technically, specialization means that the specification of the base template is conservatively extended, i.e. all formulas stated in the base template have to be valid for the specialized template.

  - *Aggregation* describes a *composite object* which is composed from several base instances. Again, all properties of the incorporated components have to be valid in the aggregation, i.e. we may refer to local properties in the context of the aggregation.

- All instances in an *object system* are related by at least one of the mechanisms described above. This implies the following:

  - Semantically, an object system is considered to be an aggregation incorporating all instances.

  - Underlying an object system we have *global time*. Concurrency is either modeled by interleaving or by the simultaneous occurrence of events at a single time instant.

4 Semantic Concepts

This chapter describes semantic foundations for our approach. We start by introducing the identification and referencing concepts underlying TROLL. After that, the basic definitions for the data universe underlying TROLL specifications are given. Finally, we introduce a logical framework called *Object Specification Logic* by Amilcar Sernadas, Cristina Sernadas, and Felix Costa [SSC92]. This framework will be used to explain the semantics of language constructs of TROLL. We will not go through the logic in full detail but we will repeat only the important definitions. For a more complete description we refer the reader to [SSC92].

4.1 Identifiers and Keys in TROLL

Based on the concepts discussed in Section 3.5.2, let us now describe the solution we envisage for object identification and referencing in TROLL.

When we used the previous version of the language (TROLL91) in examples, we were using the construction mechanism for identification spaces in the same way we would use keys in data modeling. The semantics, however, was very much different: We defined a constructed data sort of type **tuple** over a number of other sorts. The elements of this sort were the identifiers of instances in the sense of logical internal names. In this respect we were very close to approaches like FOOPS [GM87] or Glider [DDRW91].

The selectors of the tuple sort (also called projections) were used in TROLL91 like usual attributes. The major drawback is that identifying attributes in TROLL91 are *constant!* This is not always feasible in an application.

Thus, in the version TROLL described in this book the solution is the following:

- Each object instance is uniquely identified by an element of an *identifier sort*. Such an element is a *data value* of which, however, we do not know the representation—it is a value of an *abstract data type*. An identifier sort corresponds to a class in FOOPS or a type in Glider. Moreover, an element of an identifier sort (an *identifier*) corresponds to what Kent calls a token [Ken91].

- Identifier sorts are ordered, i.e. we are able to specify that an identifier sort is a subsort of another. The subsort relationship defines an inclusion on the carrier sets of the sorts, i.e. each value of the subsort is also a value of the supersort, but not vice versa. This allows us to handle different aspects of the same conceptual object (specialization).

- Since an identifier is a data value and an identifier sort can be specified by a data type specification, we can construct *terms* of identifier sorts. Any term of the proper sort *denotes* an identifier in the algebraic sense and can thus be used as a *proper name* [WJ91] or a *symbol* [Ken91].

- Terms denoting identifiers are called *references* in our approach. References correspond to *logical names* of objects and can also be used for local identification of components.

- Each identifier sort in general can be user-defined as an ordinary data type. In particular, identifier sorts can be constructed over other sorts by *generators*. Thus, we are preserving the advantages of the approach followed in TROLL91. The construction of (structured) identifier sorts is also possible in Glider. This characteristic also makes the referencing of components of a aggregation possible.

- For object referencing, we are introducing *key attributes*. These can be used like ordinary attributes, in particular we are allowing to *change the values of key attributes*. This gives us a lot more flexibility compared to TROLL91.

- Since key attributes are changeable, we need a *state-dependent mapping* of key attribute values to identifiers. These mappings can be used in terms denoting identifiers. Terms that only use key maps are called *external names*.

- The state-dependent key maps must be associated with object collections or *classes* since referencing instances is done on the level of collections, i.e. objects can only be referenced inside a larger context.

We use the following conventions when talking about identifiers:

- An identifier sort is denoted by a name enclosed in bars, e.g. |Account| .

- The name of the identifier sort corresponds to the name of a generic concept describing a collection of instances.

The referencing of objects in the language TROLL will be discussed in Section 7.1 about referencing in collections.

We cannot present a formal framework for object identification and referencing in this section since it must be embedded in the framework for explaining TROLL. Thus, the relevant aspects of the identification concepts discussed in this section are presented where they fit into the framework and the translation of language constructs. The purpose of this section is the overall presentation of the concepts in the light of the topic.

4.2 Data Signatures and Specifications

Based on concepts from algebraic specification of data types, we will now briefly give the definitions concerning the underlying data universe.

The data universe consists of an algebra of data types. It is given in an algebraic style which has become rather standard for the specification of data structures [EM85, VL89, EGL89].

We are making frequent use of data signature elements and data terms later— thus, we are introducing the basic definitions here.

We start with the definition of a data signature which provides the alphabet of sorts and functions:

Definition 4.2.1 (*Data Signature*)
A *data signature* is a tuple $\Sigma = (S, \Omega)$ where

1. S is a set of *data sorts*, and

2. Ω is an $S^* \times S$-indexed family of sets of operation symbols.

The set $\Omega_{\epsilon,s}$ contains nullary function symbols, i.e. constants. ∎

Over a data signature, we may declare variables. Variables are typed, i.e. they are associated with a sort.

Definition 4.2.2 (*Variables*)
Let $\Sigma = (S, \Omega)$ be a data signature and X an S-indexed family of sets such that $X \cap \Omega_{\epsilon} = \emptyset$. Then X is called the *set of variables over* Σ. A set X_s contains variables of sort s. ∎

An *extended signature* $\Sigma(X)$ is a data signature where variables are regarded to be nullary function symbols.

The interpretation structures for extended signatures are Σ-algebras. They associate with each sort in a signature a *carrier set* of concrete data values and with each function symbol a concrete *function* over the carrier sets of the parameters.

Definition 4.2.3 (Σ-*Algebra*)
A Σ-*Algebra* is an algebra $A(\Sigma) = (A(S), A(\Omega))$ such that

1. $A(S)$ is an S-indexed family of sets, and

2. $A(\Omega)$ is an $S^* \times S$-indexed family of functions over $A(S)$. ∎

From the symbols defined by signatures and variable declarations, we may construct *terms*. Terms denote a value which is in the carrier set of a particular sort and thus are of the same sort as the value they denote. Terms are constructed from variables and function symbols.

Definition 4.2.4 (*Data Terms*)
Let $\Sigma(X)$ be an extended signature. The set of *terms* $T_\Sigma(X)$ is defined recursively as follows:

1. Each variable $x \in X_s$ is a term of sort s.

2. If $t_1, \ldots, t_n$ are terms of sort $s_1, \ldots, s_n$, and $f: s_1, \ldots, s_n \to s$ is a function symbol in Σ, then $f(t_1, \ldots, t_n)$ is a term of sort s.

3. Nothing else is a term in $T_\Sigma(X)$. ∎

Terms are evaluated over a given interpretation, i.e. a term is associated with a particular value which is the denotation of a term.

To define the evaluation of terms w.r.t. a given interpretation $A(\Sigma)$, we have to define the *substitution* of a variable $x \in X_s$ by a value $\vartheta(x) \in A(s)$.

Definition 4.2.5 (*Substitution*)
A *substitution* ϑ of variables X w.r.t. an interpretation $A(\Sigma)$ is a family of mappings

$$\vartheta: X_s \to A(s)$$

assigning to each variable $x \in X_s$ a value from the sort carrier $A(s)$. The set of all substitutions for given $A(\Sigma)$ and X is denoted as $\Theta(A(\Sigma), X)$. ∎

The denotation of a term is evaluated recursively based on a current substitution of variables as defined in the following definition.

Definition 4.2.6 (*Term Evaluation*)
For a given interpretation $A(\Sigma)$ and a given substitution $\vartheta \in \Theta(A(\Sigma), X)$, the *valuation* function of a term is defined by

1. $(A(\Sigma), \vartheta)[x] = \vartheta(x)$ for $x \in X$

2. $(A(\Sigma), \vartheta)[f(t_1, \ldots, t_n)] = A(f)((A(\Sigma), \vartheta)[t_1], \ldots, (A(\Sigma), \vartheta)[t_n])$ ∎

Usually, we can assume a fixed interpretation structure to be given when terms are evaluated. Therefore, the denotation of a term t can be written shortly as $[t]$.

So far, we have only sketched the specification of simple sorts which are not constructed over other sorts. Let us now give briefly the definition of such *constructed data types*.

Definition 4.2.7 (*Data Type Constructors*)
Let $s, s_1, \ldots, s_n$ be data sorts with carrier sets $A(s), A(s_1), \ldots, A(s_n)$, respectively. Then

> **set**(s),
>
> **list**(s), and
>
> **tuple**$(s_1, \ldots, s_n)$

are *constructed data sorts*. Their carrier sets are

> $2^{A(s)}$ for **set**(s),
>
> $A(s)^*$ for **list**(s), and
>
> $A(s_1) \times \ldots \times A(s_n)$ for **tuple**$(s_1, \ldots, s_n)$. ∎

Data type constructors induce a number of operations which are briefly listed in Table 4.1.

As a convention, we may use *enumeration types* where we enumerate a number of constant symbols which denote the only possible values of such a type. Syntactically, an enumeration type is specified as

$$\{c_1, c_2, \ldots, c_k\}$$

where the c_i are constant symbols. On enumeration types, we do not allow other operations than the check on equality.

Data types are assumed to be specified in the usual way and will not change for a TROLL-specification and a corresponding object system. This will be called the *data universe* underlying a TROLL-specification.

Operation	Arity	Description
emptyset	$\rightarrow$ **set**(s)	set constructor
in _ _	$s\times$ **set**$(s)\rightarrow$ bool	element of?
include _ _	$s\times$ **set**$(s)\rightarrow$ **set**(s)	put element into set
remove _ _	$s\times$ **set**$(s)\rightarrow$ **set**(s)	remove element from set
card _	**set**$(s)\rightarrow$ nat	cardinality
emptylist	$\rightarrow$ **list**(s)	list constructor
in _ _	$s\times$ **list**$(s)\rightarrow$ bool	element of?
pos _ _	$s\times$ **list**$(s)\rightarrow$ nat	first position of element in list
_ **[**_**]**	**list**$(s)\times$ nat$\rightarrow s$	selection
append _ _	$s\times$ **list**$(s)\rightarrow$ **list**(s)	append element to list
remove _ _	$s\times$ **list**$(s)\rightarrow$ **list**(s)	remove element from list
delete _ _	nat$\times$ **list**$(s)\rightarrow$ **list**(s)	delete i-th element
len _	**list**$(s)\rightarrow$ nat	length
π_1 _	**tuple**$(s_1,...,s_n)\rightarrow s_1$	projection
$\vdots$	$\vdots$	$\vdots$
π_n _	**tuple**$(s_1,...,s_n)\rightarrow s_n$	projection
tuple _...._	$s_1\times...\times s_n\rightarrow$ **tuple**$(s_1,...,s_n)$	tuple constructor

Table 4.1: Operation Symbols for Data Type Constructors

4.3 Object Specification Logic

The *Object Specification Logic* (OSL) as presented in [SSC92] will serve to explain the semantics of TROLL. We assume a fixed data universe be given as outlined in the previous section. We will frequently refer to the *set of all data sorts in the data universe* which is denoted by S. S includes the data sorts as well as the the identifier sorts. The presentation of the logic is by no means complete. We will restrict ourselves to the presentaion of the necessary definitions that are used later when we describe the translation of TROLL into the logic. For a more elaborated presentation of OSL please refer to [SSC92].

Underlying an OSL specification we assume a *data universe* to be predefined. In the data universe, D denotes the set of data sorts which are not identifier sorts.

4.3.1 Local Specification

Let us describe the concepts for describing the local properties of (generic) instances firstly. OSL supports the specification of local aspects as well as *local reasoning*. In the following subsections we will present the most important concepts.

4.3.1.1 Local Logical Language

The local language defines the symbols and rules for building local specifications. The variant symbols of a local language are defined by a *signature*. Let S be the set of all data sorts defined in that data universe. In particular, *identifier sorts* are elements in S.

Possible identifier sorts are:

- *Atomic identifier sorts* which are simple sorts; and

- *Product identifier sorts* which are of the form

$$is = \mathbf{tuple}(s_1, s_2, ..., s_n)$$

For a product identifier sort is, we have projection functions $\pi_1\colon is \to s_1$, $\pi_2\colon is \to s_2, ... \pi_n\colon is \to s_n$.

Atomic identifier sorts are partially ordered, i.e. we have a relation $\leq$ defined over them.

A local signature declares the symbols that can be used in a specification.

Definition 4.3.1 (*Local Signature*)
A *local signature* is a tuple $\Theta = (\mathcal{IS}, \alpha, ATT, EVT)$ where

- $\mathcal{IS} = (IS, \leq)$, where $IS \subseteq S$, is a partially ordered set of (identifier) sorts;

- α is a sort in IS (*local sort*)

- ATT is an $S^* \times S$-indexed family of finite sets (of *Attribute Symbols*) and

- EVT is an S^*-indexed family of finite sets (of *Event Symbols*). ∎

In OSL, a signature defines first of all the identifier sorts that are introduced and used in a local specification. Recall that the convention we made for the notation of identifier sorts was to enclose the sort name in bars. The *local sort* is the identifier sort for instances of the local specification.

Please note that the set of identifier sorts may include also *products* over atomic identifier sorts. These denote aggregations which are explained in Section 4.3.2. For aggregation template specifications the local sort is such a product sort.

A symbol $a \in ATT_{w,s}$ (where $w \in S^*$) denotes an attribute in the sense of data modeling: it stands for an observable property. Note that for attributes in the usual sense which only have a value the parameter list w is the empty list. Attribute symbols with parameters are abbreviations for sets of attributes indexed by the parameter sorts. Each attribute must have a *codomain s* which provides the sort of possible values of the attribute.

A symbol $e \in EVT_w$, where $w \in S^*$, is an event symbol. Event symbols denote (sets of) events which are the basic state transitions in our approach. An event symbol may have a (possibly empty) list of parameter sorts w. A parameterized event symbol denotes an indexed set of events.

Let us give an example to illustrate possible signatures for local specifications. Consider the specification of an *account*. The observations we would like to be able to do over an account are the current balance in terms of a value of a (user-defined) data type **money**, a reference to a holder (in terms of an identifier type |Customer|), and a number of type **nat**. Thus, the $\mathcal{IS}$ component of a local signatue Θ_{Account} would be declared as follows:

$$IS_{\text{Account}} = \{|\text{Account}|, |\text{Customer}|\}$$

The partial order relation on IS is the identity. We are declaring |Account| to be the local sort for the local specification.

$$\alpha_{\text{Account}} = |\text{Account}|$$

The attributes ATT_{Account} of Θ_{Account} are the following:

$$(ATT_{\text{Account}})_{\text{nat}} = \{\text{Number}\}$$
$$(ATT_{\text{Account}})_{\text{money}} = \{\text{Balance}\}$$
$$(ATT_{\text{Account}})_{|\text{Customer}|} = \{\text{Holder}\}$$

As local events that can happen to an account we consider the following: accounts can be opened and closed; thus, we have to model these by events. Furthermore, we may deposit money (*credit*) and withdraw money (*debit*). Let the family of event symbols EVT_{Account} in Θ_{Account} be defined as

$$(EVT_{\text{Account}})_{\text{nat},|\text{Customer}|} = \{\text{open}\}$$
$$(EVT_{\text{Account}})_{\epsilon} = \{\text{close}\}$$
$$(EVT_{\text{Account}})_{\text{money}} = \{\text{credit}, \text{debit}\}$$

Over a local signature Θ we define a set of local predicate symbols. These are the basic building blocks of formulas that describe properties of instances.

Definition 4.3.2 (*Local Predicate Symbols*)
The S^*-indexed family of sets of *local predicate symbols* $\mathcal{P}(\Theta)$ over a local signature $\Theta = (\mathcal{IS}, \alpha, ATT, EVT)$ is defined as follows:

1. $\{\rhd\, e \mid e \in EVT_w\} \in \mathcal{P}(\Theta)_w$ where $w \in S^*$

2. $\{\rhd\, a \mid a \in ATT_{w,s}\} \in \mathcal{P}(\Theta)_{w,s}$ where $w \in S^*$ and $s \in S$

3. $\{\odot\, e \mid e \in EVT_w\} \in \mathcal{P}(\Theta)_w$ where $w \in S^*$ ∎

A predicate symbol $\rhd\, e$ for $e \in EVT$ indicates that an event e is *enabled* (in the current state), i.e. it may occur, but does not have to. A predicate symbol $\rhd\, a$ for $a \in ATT$ should be interpreted as "a certain value of a is *observable*". This approach allows to handle (temporary) *undefined attributes* very nicely: when we cannot observe anything for an attribute, it is undefined. A predicate symbol $\odot\, e$ for $e \in EVT$ refers to the *occurrence* of an event in the current state. Here we depart a bit from the original proposal in that we do not regard the reading of an attribute value as an event.

Over the predicates and over data terms we define the *local template language* of which the formulas of a template specification are sentences.

Definition 4.3.3 (*Local Template Language*)
Let Θ be a template signature and X be a set of variables over S. The *local template language* $\mathcal{TL}_\Theta(X)$ is defined inductively as follows:

1. $t_1 = t_2 \in \mathcal{TL}_\Theta(X)$ if t_1 and t_2 are data terms of the same data sort s;

2. $p(t_1, ..., t_n) \in \mathcal{TL}_\Theta(X)$ if $p \in \mathcal{P}(\Theta)_{s_1,...s_n}$ and each t_i is a term of sort s_i;

3. $(\neg\, \varphi) \in \mathcal{TL}_\Theta(X)$ if $\varphi \in \mathcal{TL}_\Theta(X)$;

4. $(\varphi \Rightarrow \psi) \in \mathcal{TL}_\Theta(X)$ if $\varphi, \psi \in \mathcal{TL}_\Theta(X)$;

5. $((\forall x{:}\, s)\varphi) \in \mathcal{TL}_\Theta(X)$ if $\varphi \in \mathcal{TL}_\Theta(X)$ and $x \in X$;

6. $(\bigcirc \varphi), (\Box \varphi)$ if $\varphi \in \mathcal{TL}_\Theta(X)$;

7. $(\varphi\ \mathcal{U}\ \psi) \in \mathcal{TL}_\Theta(X)$ if $\varphi, \psi \in \mathcal{TL}_\Theta(X)$. ∎

For convenience, let us immediately introduce derived operators which are used in OSL specifications:

- $(\varphi \vee \psi) \equiv ((\neg\varphi) \Rightarrow \psi)$

- $(\varphi \wedge \psi) \equiv (\neg(\varphi \Rightarrow (\neg\psi)))$

- $(\varphi \mid \psi) \equiv ((\varphi \wedge (\neg\psi)) \vee ((\neg\varphi) \wedge \psi))$

- $(\varphi \Leftrightarrow \psi) \equiv ((\varphi \Rightarrow \psi) \wedge (\psi \Rightarrow \varphi))$

- $(\Diamond\varphi) \equiv (\neg(\Box(\neg\varphi)))$

- $(\exists x\colon s)\varphi \equiv (\neg(\forall x\colon s)\neg\varphi)$

Like in first order logic with equality, equality predicates are the basic atomic formulas. Furthermore, special predicates in OSL as defined above (2.) are atomic formulas in OSL. The first-order logical connectives $\neg$ for negation, $\wedge$ for conjunction, $\vee$ for disjunction, $|$ for exclusive disjunction, $\Rightarrow$ for implication, and $\Leftrightarrow$ for equivalence are provided in OSL. For the temporal conncetives we used the notation of Manna and Pnueli [MP92a]: $\bigcirc$ is the nexttime operator, $\Diamond$ is the eventually operator, $\Box$ is the always operator, and $\mathcal{U}$ is the until operator. The intuitive meaning of these operators is the following:

- A formula $(\bigcirc\varphi)$ currently holds if φ holds in the next state.

- A formula $(\Diamond\varphi)$ currently holds if φ holds in the current state or in some future state.

- A formula $(\Box\varphi)$ currently holds if φ holds in the current state and in every future state.

- A formula $(\varphi\ \mathcal{U}\ \psi)$ currently holds if φ holds in the current or some future state and ψ holds at least in every state until φ holds.

The formal semantics of these operators is defined in Definition 4.3.9 in the next section.

A special operator that will be frequently used later is the *unless* operator. It is defined based on the *until* operator:

$$(\varphi\ \mathcal{W}\ \psi) \equiv ((\varphi\ \mathcal{U}\ \psi) \vee (\Box\varphi))$$

The difference to the until-operator is that ψ is not required to become valid.

Please note that the exclusive-or operator $|$ is not associative. Thus, we use an operator $\big|$ which is defined as

$$\big|\,(\varphi_1, ..., \varphi_n) \equiv \begin{array}{l} \varphi_1 \wedge \neg\varphi_2 \wedge ... \wedge \neg\varphi_n \\ \vee \neg\varphi_1 \wedge \varphi_2 \wedge \neg... \wedge \neg\varphi_n \\ \vdots \\ \vee \neg\varphi_1 \wedge ... \wedge \varphi_n \end{array}$$

An abbreviation concerns equations. An equation $t = \mathbf{true}$, where t is of type bool, is often written as t.

A sublanguage of $\mathcal{TL}_\Theta(X)$ contains formulas that are conjunctions or disjunctions of occurrence predicates of the form $\odot e(t_1, ..., t_n)$ where $e \in EVT_{s_1,...,s_n}$ and each t_i is a term of sort s_i. These formulas are called *interaction formulas*. We may express the simultaneous occurrence of local events as well as the nondeterministic choice between local events.

A characteristic feature of OSL is that it allows to analyze specifications of components in isolation. Therefore, we have to be able to associate formulas with a particular instance, i.e. to *localize* formulas. This localization is achieved by binding predicates to identifiers—we obtain *local predicates* this way. In the following definition, we present the language of localized formulas which we call *local specification language*.

Definition 4.3.4 (*Local Specification Language*)
Let $\Theta = (\mathcal{IS}, \alpha, ATT, EVT)$ be a local signature. A pair $(X, i(\varphi))$ is a sentence of the *local specification language*$\mathcal{TSL}_\Theta$ iff

1. X is a set of sorted variables over S such that $i \in X$;

2. $i \in X_{\alpha(\Theta)}$, i.e. i is an identifier;

3. $\varphi \in \mathcal{TL}_\Theta(X)$ is a formula of the template language over Θ and X;

4. φ does not contain any quantification on the variable i and

5. $i(\varphi)$ is obtained from φ by replacing every predicate symbol $p \in \mathcal{P}(\Theta)$ in φ by $i.p$. ∎

A local specification is composed from a local signature and a set of local specification formulas:

Definition 4.3.5 (*Local Specification*)
A *local specification* is a pair $\mathbf{T} = (\Theta, Ax)$ where Θ is a template signature and $Ax \subseteq \mathcal{TSL}_\Theta$. The elements of Ax are called *local axioms*. ∎

In the sequel, we will only use formulas of the local template language $\mathcal{TL}_\Theta(X)$ since we specify generic instances.

4.3.1.2   Local Semantics

The interpretation structures for local specifications are built over an interpretation structure for the data universe as depicted above in Section 4.2. As usual, $A(s)$ denotes a carrier set for a sort s.

The intended interpretation structures for linear temporal logic are usually *sequences of states*. Using temporal logic formulas, we are able to talk about the validity of assertions in certain states relative to the current state. First of all, it is therefore important to define the *possible states*. A state is described by a set of predicates that hold in that particular state. Please note that we are talking about *local states*: each valid predicate is a local predicate.

Definition 4.3.6 (*Possible Local States*)
Let Θ be a local signature. The set of *possible local states* S_Θ over Θ is defined as

$$S_\Theta = 2^\Pi$$

where

$$\Pi = \{p(x_1, ..., x_n) \mid p \in \mathcal{P}(\Theta)_{s_1,...,s_n},\ x_i \in A(s_i)\ \textit{for } i = 1, ..., n\}$$

An element of S_Θ is called a *local state* and is notated as σ.   ■

In a local interpretation structure, we are defining a *generic instance* in terms of a sequence of possible states. Thus, we are associating with points in time sets of valid predicates that characterize the state of an instance at that particular point in time. The semantics of template specification formulas is defined over such interpretation structures.

Definition 4.3.7 (*Local Interpretation Structure*)
Let Θ be a template signature. An *interpretation structure* over Θ is a pair $\mathbf{I} = (A(IS), \hat{\sigma})$ where

- $A(IS)$ is a family of carrier sets for the identifier sorts in $\mathcal{IS}$;

- $\hat{\sigma} = \sigma_0 \sigma_1 ...$ is a sequence of states associating with each instant in time a state $\sigma_k \in S_\Theta$, $k \in \mathbb{N}_0$, such that the following holds:

  1. if $(\odot\, e(x_1, ..., x_n)) \in \sigma_k$ then $(\triangleright e(x_1, ..., x_n)) \in \sigma_k$;

  2. if
  $$\{\triangleright a \mid (\triangleright a) \in \sigma_{k+1}\} \neq \{\triangleright a \mid (\triangleright a) \in \sigma_k\} \quad \text{or}$$
  $$\{\triangleright e \mid (\triangleright e) \in \sigma_{k+1}\} \neq \{\triangleright e \mid (\triangleright e) \in \sigma_k\}$$

  then there is an event symbol e such that $\odot\, e(x_1, ..., x_n) \in \sigma_k$ for suitable $x_1, ..., x_n$;

3. for each attribute symbol $a \in ATT$ the following holds: if

$$(\triangleright a(x_1, ..., x_m, x)) \in \sigma_k \text{ and } (\triangleright a(x'_1, ..., x'_m, x')) \in \sigma_k$$
$$\text{and } x_i = x'_i \text{ for } i = 1, ..., m$$

then $x = x'$.

4. for each event symbol $e \in EVT$ the following holds: if

$$(\odot e(x_1, ..., x_m)) \in \sigma_k \text{ and } (\odot e(x'_1, ..., x'_m)) \in \sigma_k$$

then $x_i = x'_i$ for $i = 1, ..., m$. ∎

This definition states the following for interpretation structures or generic instances:

1. The first requirement demands that the occurrence of an event in state k implies that the event is enabled in state k. In other words this means that a event can only occur in a state if it is enabled in that state.

2. Requirement 2. demands that solely the occurrence of an event can change the set of attribute observations and the set of enabled events (*Frame Assumption*).

3. The next requirement demands that attributes be functional—in an given situation, only one value for an attribute may be observed.

4. The last requirement describes the mutual exclusion of events of the same kind—no two instantiations of the same event symbol may occur in parallel in the same state.

In order to be able to interpret first-order formulas, we have to assign values to variables:

Definition 4.3.8 (*Assignment*)
Let **I** be an interpretation structure over Θ and let X be an S-indexed family of variables. An *assignment* ϑ is an S-indexed family of maps $\vartheta_s\colon X_s \to A(S)$ associating a value $v \in A(s)$ with a variable $x \in X_s$. ∎

The interpretation of terms and variables is the usual one described in the previous section about data signatures and specifications. Now we may define the semantics of template formulas over local interpretation structures:

Definition 4.3.9 (*Semantics of Local Formulas*)
Let $\mathbf{I}$ be an interpretation structure over a template signature Θ, X an S-indexed family of variables, ϑ an assignment to X and $k \in I\!N_0$ a position. The *semantics* of formulas over X is defined as follows:

$$\mathbf{I}[\vartheta, k] \models p(t_1, ..., t_n) \qquad \text{iff } p([\![t_1]\!], ..., [\![t_n]\!]) \in \sigma_k \text{ for } p \in \mathcal{P}(\Theta)$$

$$\mathbf{I}[\vartheta, k] \models (\neg\varphi) \qquad \text{iff not } \mathbf{I}[\vartheta, k] \models \varphi$$

$$\mathbf{I}[\vartheta, k] \models (\varphi \Rightarrow \psi) \qquad \text{iff } \mathbf{I}[\vartheta, k] \models (\varphi) \text{ implies } \mathbf{I}[\vartheta, k] \models (\psi)$$

$$\mathbf{I}[\vartheta, k] \models ((\forall x\colon s)\varphi) \qquad \text{iff } \mathbf{I}[\vartheta', k] \models (\varphi) \text{ for any assignment } \vartheta' \text{ x-equivalent to } \vartheta$$

$$\mathbf{I}[\vartheta, k] \models (\bigcirc\,\varphi) \qquad \text{iff } \mathbf{I}[\vartheta, k + 1] \models (\varphi)$$

$$\mathbf{I}[\vartheta, k] \models (\Box\varphi) \qquad \text{iff } \mathbf{I}[\vartheta, k'] \models (\varphi) \text{ for every } k' \geq k$$

$$\mathbf{I}[\vartheta, k] \models (\varphi\,\mathcal{U}\,\psi) \qquad \text{iff } \mathbf{I}[\vartheta, k'] \models (\psi) \text{ for some } k' \geq k \text{ and } \mathbf{I}[\vartheta, k''] \models (\varphi) \text{ for every } k'' \text{ such that } k \leq k'' < k'$$

An assignment ϑ' is called *x-equivalent* to an assignment ϑ if for each variable $y \neq x$ holds $\vartheta(y) = \vartheta'(y)$.

We now define the notions of *satisfaction* and *validity* of formulas.

Definition 4.3.10 (*Satisfaction of Template Formulas*)
Let Θ be a template signature, X be a set of variables, and $\mathbf{I}$ be an interpretation structure over Θ. $\mathbf{I}$ *satisfies a formula* $\varphi \in \mathcal{TL}_\Theta$, written $\mathbf{I} \models \varphi$, if $\mathbf{I}[\vartheta, k] \models \varphi$ for every assignment ϑ and every position $k \in I\!N_0$.

Definition 4.3.11 (*Validity*)
Let Θ be a local signature. A formula $\varphi \in \mathcal{TL}_\Theta$ is *valid* if $\mathbf{I} \models \varphi$ for every interpretation structure $\mathbf{I}$ over Θ.

The semantic closure of a template specification includes all template specification formulas that are valid for each interpretation structure $\mathbf{I}$ of Θ.

Definition 4.3.12 (*Semantic Closure*)
Let $\mathbf{T} = (\Theta, Ax)$ be a template specification. The *semantic closure* of $\mathbf{T}$ is the following set of formulas:

$$Ax^{\models\Theta} = \{(\varphi) \mid \mathbf{I} \models (\varphi) \text{ for each interpretation structure } \mathbf{I} \text{ over } \Theta \text{ such that } \mathbf{I} \models (\psi) \text{ for each } \psi \in Ax\}$$

We write $\models_{\mathbf{T}} (\varphi)$ if $(\varphi) \in Ax^{\models\Theta}$.

Definition 4.3.13 (*Entailment*)

Let $\mathbf{T} = (\Theta, Ax)$ be a template specification and H be a set of formulas in $\mathcal{TSL}_\Theta$. The set of formulas *semantically entailed* by H within $\mathbf{T}$ is defined as follows:

$$Ax^{H\models\Theta} = \{(\varphi) \mid \ \mathbf{I} \models (\varphi) \text{ for each interpretation structure } \mathbf{I} \text{ over } \Theta$$
$$\text{such that } \mathbf{I} \models \psi \text{ for all } \psi \in (Ax \cup H)\}$$

We write $H \models (\varphi)$ if $(\varphi) \in Ax^{H\models\Theta}$. ∎

4.3.1.3 Local Reasoning

A central feature of a logical approach is the presence of a *deductive system* which allows us to derive knowledge from a given specification. Usually, a deductive system consists of

- a set of *axioms*, and

- a set of *rules*, which provide a syntactical means to derive valid formulas from formulas that have already been proven to be valid.

In OSL, the approach is a bit different. The logic is given by the set of formulas that is generated from a specification and which is closed against a collection of *inference rules*. The inference rules are the traditional ones for many-sorted, first-order, linear temporal logic with equality along with some specific rules related to OSL properties. We will not present the entire set of inference rules here but restrict ourselves to the most interesting ones concering temporal connectives and OSL specific properties. The entire set of inference rules can be found in [SSC92]. For a light introduction into temporal reasoning in many-sorted, first-order, linear temporal logic see [MP92a, Sections 3.5–3.10].

Local Inference Rules

For convenience we use the notation $\vec{x}$ for a vector $x_1, ..., x_k$ of variables.

$$(\Box(\varphi \Rightarrow \psi)) \Rightarrow ((\Box\varphi) \Rightarrow (\Box\psi)) \in Ax^{\vdash\Theta} \qquad (Loc\Box)$$

This rule states that distributing a $\Box$ operator over an implication may only weaken it. If $\varphi \Rightarrow \psi$ holds at all future positions, and φ holds at all future positions, then also ψ holds always in the future.

$$((\bigcirc(\varphi \Rightarrow \psi)) \Rightarrow ((\bigcirc\varphi) \Rightarrow (\bigcirc\psi))) \in Ax^{\vdash\Theta} \qquad (Loc\bigcirc)$$

This rule says that $\bigcirc$ distributes over an implication: $\varphi \Rightarrow \psi$ holds in the next position only if $(\bigcirc \varphi) \Rightarrow (\bigcirc \psi)$ holds in the current position.

$$((\bigcirc(\neg\varphi)) \Leftrightarrow (\neg(\bigcirc\varphi))) \in Ax^{\vdash\ominus} \qquad\qquad (Loc\neg\bigcirc)$$

The rule states the duality of the $\bigcirc$ operator.

$$((\Box\varphi) \Leftrightarrow (\varphi \wedge (\bigcirc(\Box\varphi)))) \in Ax^{\vdash\ominus} \qquad\qquad (Loc\Box Exp)$$

This rule states that φ holds always in the future iff it holds in the current state and it always holds from the next state on.

$$((\Box(\varphi \Rightarrow (\bigcirc\varphi))) \Rightarrow (\varphi \Rightarrow (\Box\varphi))) \in Ax^{\vdash\ominus} \qquad\qquad (LocInd)$$

This rule is an induction rule. It states that if it is always in the future the case that if φ holds in the current state it holds in the next state then whenever φ holds at a position, it holds at all subsequent positions.

$$((\varphi\,\mathcal{U}\,\psi) \Rightarrow (\Diamond\psi)) \in Ax^{\vdash\ominus} \qquad\qquad (LocUnt)$$

This rule directly reflects the definition of the $\mathcal{U}$ operator. If $\varphi\,\mathcal{U}\,\psi$ holds in the current state then eventually ψ must hold.

Let us now turn to rules that reflect specific properties of OSL. Basically, they reflect the requirements that were stated for interpretation structures of template signatures.

$$(\odot e(\vec{x}) \Rightarrow \rhd e(\vec{x})) \in Ax^{\vdash\ominus} \text{ for each event symbol } e \in EVT_\Theta \qquad (Loc\odot\rhd)$$

This rule demands that a predicate denoting the occurrence of an event in a state can only be valid if the event is enabled in that state.

$$(\, \exists a \in ATT_\Theta \; \exists \vec{x}\colon \neg(\rhd a(\vec{x}) \Leftrightarrow (\bigcirc\rhd a(\vec{x}))) \qquad\qquad (LocAttFrm)$$
$$\Rightarrow ((\exists e \in EVT_\Theta \; \exists \vec{y}\colon \odot e(\vec{y}))\,) \in Ax^{\vdash\ominus}$$

This rule states that if the set of observations changes from the current state to the next, then there must be an event that occurs in the current state.

$$\left((\rhd a(\vec{x},y) \wedge \rhd a(\vec{x'},y') \wedge (\vec{x} = \vec{x'})) \Rightarrow (y = y')\right) \in Ax^{\vdash\ominus} \qquad (LocFunc)$$
$$\text{for each attribute symbol } a \in ATT_\Theta$$

This rules reflects the requirement that attributes be functional which was stated for interpretation structures of template signatures.

$$\left((\odot e(\vec{x}) \wedge \odot e(\vec{x'})) \Rightarrow (\vec{x} = \vec{x'})\right) \in Ax^{\vdash_\Theta} \qquad\qquad (LocExcl)$$

$$\text{for each event symbol } e \in EVT_\Theta$$

This rule reflects the requirement that different instantiations of the same event symbol cannot occur in the same state.

In the next definition, we define a theory as being a set of formulas over a signature which is closed against the inference rules.

Definition 4.3.14 (*Local Theory*)
Let $\mathbf{T} = (\Theta, Ax)$ be a template specification. The *local theory* generated by $\mathbf{T}$, written $Ax^{\vdash_\Theta}$, is the least set of formulas in $\mathcal{TL}_\Theta$ satisfying the inference rules. Each element (φ) of $Ax^{\vdash_\Theta}$ is called a *theorem* of $\mathbf{T}$, written $\vdash_{\mathbf{T}} (\varphi)$. ∎

It has been shown in [SSC92] that the inference system presented in that paper is sound. Completeness has not been proven yet.

Definition 4.3.15 (*Derived Formulas*)Let $\mathbf{T} = (\Theta, Ax)$ be a template specification and $H \subseteq \mathcal{TL}_\Theta$. The set of formulas *derived from* H, written $(Ax \cup H)^{\vdash_\Theta}$, is the least set satisfying the following requirements:

1. $Ax^{\vdash_\Theta} \subseteq (Ax \cup H)^{\vdash_\Theta}$

2. $H \subseteq (Ax \cup H)^{\vdash_\Theta}$

3. $\left[(\varphi \Rightarrow \psi) \in (Ax \cup H)^{\vdash_\Theta} \wedge (\varphi) \in (Ax \cup H)^{\vdash_\Theta}\right] \Rightarrow (\psi) \in (Ax \cup H)^{\vdash_\Theta}$

We write $H \vdash_{\mathbf{T}} (\varphi)$ whenever $(\varphi) \in (Ax \cup H)^{\vdash_\Theta}$. ∎

Proposition 4.3.16 Let $\mathbf{T} = (\Theta, Ax)$ be a template specification and $(\varphi) \in \mathcal{TL}_\Theta$ and $(\psi) \in \mathcal{TL}_\Theta$ be template formulas over Θ. Then the following propositions hold:

1. $(\varphi) \vdash_{\mathbf{T}} (\psi) \Rightarrow \vdash_{\mathbf{T}} ((\varphi) \Rightarrow (\psi))$ $\qquad\qquad (DED)$

2. $[((\varphi) \vdash_{\mathbf{T}} (\psi)) \wedge ((\varphi) \vdash_{\mathbf{T}} (\neg\psi))] \Rightarrow \vdash_{\mathbf{T}} (\neg\varphi)$ $\qquad\qquad (ABS)$

 ∎

The first proposition *DED* says that if a formula can be deduced from another one, then an implication specifying this is a theorem (*deduction*). The second proposition says that if we can deduce from a formula (φ) another formula (ψ) and its negation $(\neg\psi)$, then the formula (φ) is not a local theorem (*reductio ad absurdum*).

4.3.2 Template Morphisms

A special feature of the OSL framework is that we may express the construction of (composed) template specifications from simpler template specifications. In OSL, this is achieved by introducing *template morphisms*. Basically, template morphisms are based on a mapping between the signature elements of the source template signature into signature elements of the target template signature. Along the mapping of the signature elements, we may translate template axioms and thus embed template specifications into others. Two types of template morphisms are distinguished:

- An *incorporation* means that the source signature becomes part of the target signature (it is *injected*). This reflects the part-of relation or aggregation. Each axiom of the source template specification is saved in the target template specification, i.e. we may prove things about each component when looking at the aggregated template, even if we have several components with the same specification.

- A *specialization* means that we may extend a signature by additional signature elements (thus *including* the source signature into the target signature). Then the axioms of the source template specification carry over without changes to the target template specification. Thus, it is a special case of an incorporation.

4.3.2.1 Signature Morphisms

The basis of a template specification morphism is a signature morphism:

Definition 4.3.17 (*Template Signature Morphism*)
Let $\Theta = (\mathcal{IS}, \alpha, ATT, EVT)$ and $\Theta' = (\mathcal{IS}', \alpha', ATT', EVT')$ be template signatures. A *template signature morphism* $\sigma \colon \Theta \to \Theta'$ is a tuple $(\sigma_{IS}, \omega_\alpha, \sigma_{ATT}, \sigma_{EVT})$ where

- σ_{IS} is a mapping of identifier sorts such that $\sigma_{IS}(s) \leq \sigma_{IS}(s')$ whenever $s \leq s'$;

- ω_α is an operation that when applied to α' yields the sort $\sigma_{IS}(\alpha)$ obtained by applying the identifier sort morphism to α;

- σ_{ATT} is an $S^* \times S$-indexed family of maps where an attribute symbol $a(s_1, ..., s_k, s)$ is mapped to $\sigma_{ATT}(a)(\sigma(s_1), ..., \sigma(s_k), \sigma(s))$;

- σ_{EVT} is an S^*-indexed family of maps where an event symbol $e(s_1, ..., s_k)$ is mapped to $\sigma_{EVT}(e)(\sigma(s_1), ..., \sigma(s_k))$. ∎

A signature morphism provides a translation of symbols into other symbols. The sorts of Θ are mapped to sorts in Θ' without violating the ordering of sorts, there is a mapping that associates the local sort α of Θ with the local sort α' of Θ', the attribute symbols are mapped to attribute symbols indexed by the corresponding sorts in $D \cup IS'$ and the event symbols are mapped to event symbols in Θ' indexed by the corresponding sorts in $D \cup IS'$.

The operation ω_α tells us how the two signatures are related:

- If ω_α is an inclusion (i.e. the initial sort α' is a subsort of the image $\sigma_S(\alpha)$ of α), we are specifying an *inclusion* of the template specification having the sort α as local sort into a template specification having the sort α' as local sort.

- If $\omega_S(\alpha)$ is obtained from α' by a projection σ_α, then we are specifying an *injection* of the template specification having the sort α as local sort into a template specification having the sort α' as local sort.

We are now going to define the two types of template signature morphism that are used in the object specification logic.

Definition 4.3.18 (*Inclusion*)
A template signature morphism $\sigma: \Theta \to \Theta'$ where $\Theta = (\mathcal{IS}, \alpha, ATT, EVT)$ and $\Theta' = (\mathcal{IS}', \alpha', ATT', EVT')$ is an *inclusion* if σ_S is an inclusion, σ_α is a sort inclusion, and $\sigma_{ATT}, \sigma_{EVT}$ are inclusions. ∎

An inclusion thus states that the signature Θ is included in the signature Θ'.

Definition 4.3.19 (*Injection*)
A template signature morphism $\sigma: \Theta \to \Theta'$ where $\Theta = (\mathcal{IS}, \alpha, ATT, EVT)$ and $\Theta' = (\mathcal{IS}', \alpha', ATT', EVT')$ is an *injection* if σ_S is an inclusion, σ_α is a sort projection, and $\sigma_{ATT}, \sigma_{EVT}$ are injective mappings. ∎

A signature morphism is the basis for a translation of formulas:

Definition 4.3.20 (*Translation of Formulas*)
Let $\sigma: \Theta \to \Theta'$ be a signature morphism. The *translation of a formula* $(\varphi) \in \mathcal{TL}_\Theta$ by σ is the formula $(\sigma(\varphi)) \in \mathcal{TL}_{\Theta'}$ such that $\sigma(\varphi) \in \mathcal{TL}_{\Theta'}$ is obtained from φ by replacing each predicate symbol $p \in \mathcal{P}_\Theta$ by its translation $\sigma(p)$. ∎

4.3.2.2   Specification Morphisms

Definition 4.3.21 (*Template Specification Morphism*)
A *template specification morphism* $\sigma\colon (\Theta, Ax) \to (\Theta', Ax')$ is a template signature
morphism such that the set of translations of axioms in Ax is included in the
semantic closure $(Ax')^{\models_{\Theta'}}$. ∎

This defintion means that a morphism between template specifications preserves
the specification of the source template specification— all axioms (in their translated form) must remain valid in the target template specification.

Now we are defining two special morphisms that constitute the basic constructors
in OSL:

Definition 4.3.22 (*Incorporation*)
A template specification morphism is an *incorporation* iff the underlying template
signature morphism is an injection. ∎

Definition 4.3.23 (*Specialization*)
A template specification morphism is a *specialization* iff the underlying template
signature morphism is an inclusion. ∎

Incorporation morphisms are essential in the presence of aggregation. In an
aggregation, we want the axioms of *each* component to be entailed once by the
aggregation template. Consider e.g. the aggregation of two account specifications.
We have two injective signature morphisms

$$\rho_1\colon \Theta_{|\mathsf{Account}|} \to \Theta_{(|\mathsf{Account}|,|\mathsf{Account}|)}$$

and

$$\rho_2\colon \Theta_{|\mathsf{Account}|} \to \Theta_{(|\mathsf{Account}|,|\mathsf{Account}|)}$$

that induce incorporations of template specifications for single accounts into a
template specification describing aggregations of two accounts. The translation of
every axiom in $Ax_{|\mathsf{Account}|}$ is entailed twice for the aggregation (it would not be
duplicated for a specialization). Looking at the aggregation, we may refer to the
properties of each component using the projections $\mu_1.\mu_2$ on the identifier sort (a
tuple sort) to obtain the local axioms. This will be illustrated further in the next
section.

4.3.3   Object System Specification

In the previous section we have shown how local specifications can be constructed
over other local specifications. Let us now turn to the specification of systems of
(interacting) objects.

4.3.3.1 Global Language

An object system consists of a collection of objects. Each object is an instance of an associated template specification. Thus, the vocabulary of a system specification consists of the signatures of all templates for the components.

Definition 4.3.24 (*System Signature*)
A *system signature* is a pair $\Psi = (\mathcal{IS}, \{\Theta^{is}\}_{is \in IS})$ such that

- $\mathcal{IS} = (IS, \leq)$ is a partially ordered set of identifier sorts such that all possible products of sorts in IS are also in IS equipped with projections $\pi_i : \mathbf{tuple}(is_1, ..., is_n) \to is_i$;

- for each $s \in IS$, $\Theta^s = (\mathcal{IS}^s, s, Att^s, EVT^s)$ is a template signature such that each $s' \in IS^s$ is an element of IS;

- if $s \leq s'$, then there is an inclusion morphism $\eta_{s',s} \colon \Theta^{s'} \to \Theta^s$;

- for each aggregation sort $s = \mathbf{tuple}(s_1, ..., s_n) \in IS$ with $s_1, ..., s_n$ *atomic*, Θ^s is the aggregation obtained by incorporating $\Theta^{s_1}, ..., \Theta^{s_n}$ using injection morphisms $\rho_1 \colon \Theta^{s_1} \to \Theta^s, ..., \rho_n \colon \Theta^{s_n} \to \Theta^s$. ∎

Definition 4.3.25 (*Interaction Language*)
Let $\Psi = (\mathcal{IS}, \{\Theta^s\}_{s \in IS})$ be a system signature. The *interaction language* $\mathcal{IL}_{\Theta_s}$ for the aggregation signature Θ_s with local sort $\alpha_{\Theta^s} = s = \mathbf{tuple}(s_1, ..., s_n)$ includes formulas of the form

1. $(\varphi \Rightarrow \psi)$ and

2. $(\psi \Rightarrow \varphi)$

where φ and ψ are local interaction formulas over template signatures Θ^{s_i} and Θ^{s_j}, $(i \neq j)$. They include occurrence predicates local to the i-th or j-th component of the aggregation. ∎

The interaction language thus allows us to describe relationships between local event occurrences of components when put into a larger context. A formula $\varphi \Rightarrow \psi$ where φ and ψ are local interaction formulas means that the occurrence of events described by φ implies the simultaneous occurrence of events described by ψ which are events of another component of the aggregation.

This is used in a *system specification* to describe interactions between the components:

Definition 4.3.26 (*System Specification*)

A *system specification* is a pair $\Gamma = (\Psi, \{Ax^s\}_{s \in IS})$ such that

- $\Psi = (\mathcal{IS}, \{\Theta^s\}_{s \in IS})$ such that for all $s \in IS$ $\mathbf{T}^s = (\Theta^s, Ax^s)$ is a template specification;

- for each aggregation $s = \mathbf{tuple}(s_1, ..., s_n)$ such that $s_1, ..., s_n$ are atomic sorts, the set of local axioms Ax^s does only include interaction formulas $Ax^s \subseteq \mathcal{IL}_{\Theta_s}$

                                                                           ■

Thus, a system specification consists of the template specifications of the system components (the *objects*) and the interaction relationships between the components which are specified as formulas over the aggregation of the interacting components.

4.3.3.2 Global Semantics

The interpretation structures of local templates (generic objects) are the components of an interpretation structures for a system signature. In the definition of a system interpretation structure we must state how the local interpretation structures are related.

Definition 4.3.27 (*System Interpretation Structure*)

Let $\Psi = (\mathcal{IS}, \{\Theta^s\}_{s \in IS})$ be a system signature with local template signatures $\Theta^s = (\mathcal{IS}^s, \alpha^s, Att^s, EVT^s)$ for each $s \in IS$. A *system interpretation structure* over Ψ is a tuple $\mathbf{SYS} = (A(IS), \{\mathbf{I}^s\}_{s \in IS})$ where

- $A(IS)$ is a family of carrier sets $A(s)$ for each $s \in IS$ where $A(s) \subseteq A(s')$ whenever $s \leq s'$;

- each $\mathbf{I}^s = (A(IS)^s, \hat{\sigma}^s)$ is a local interpretation structure for the local signature Θ^s such that the following conditions hold:

  1. $A(s')^s = A(s')$ for each $s' \in IS^s$, i.e. the interpretation of the local identifier sorts does not change when interpreted in the system context;

  2. For each specialization $\eta_{s',s}$ a state of the specialization restricted to the signature of the base must be equal to the image of the corresponding state of the base: $\hat{\sigma}_k^s \downarrow_R = \eta_{s',s}(\hat{\sigma}_k^{s'})$ where $R = \eta_{s',s}(\mathcal{S}_\Theta^{s'})$

  3. For each incorporation $\rho_i : \Theta^{s_i} \rightarrow \Theta^{(\pi_1 : s_1, ..., \pi_n : s_n)}$ a state of the aggregation restricted to the image of a component signature must be equal to the translation of the corresponding component state:

$$\hat{\sigma}_k^{is} \downarrow_R = \mu_i(\mathcal{V}^{s_i}(k))$$

  where $is = \mathbf{tuple}(s_1, ..., s_n)$ and $R = \mu_i(\mathcal{S}_\Theta^{s_i})$.   ■

In [SSC92] the following propositions have been stated for the various forms of inheritance in OSL:

- For each pair s, s' of atomic identifier sorts in IS, if $s \leq s'$ then the local specification $\eta_{s',s}$: $(\Theta^{s'}, Ax^{s'}) \rightarrow (\Theta^s, Ax^s)$ is a *specialization*.

- For each identifier sort $s = (s_1, ..., s_n)$ where $s_1, ..., s_n$ are atomic identifier sorts, then the specification morphism ρ_i: $(\Theta^{s_i}, Ax^{s_i}) \rightarrow (\Theta^s, Ax^s)$ is an *incorporation*.

4.3.3.3 Global Reasoning

As with local inference rules, we will not enumerate all inference rules in this section. Moreover, we again will stick to the most important ones and refer the reader to [SSC92] for details. The important rules are the ones that take into account *locality of components*, *specialization*, and *incorporation*.

$$(\varphi) \in Ax^{\vdash_{\Psi,s}} \text{ whenever } (\varphi) \text{ is a local theorem of } \mathbf{T}^s = (\Theta^s, Ax^s) \qquad (Loc)$$

This rule simply states that whenever a formula is a theorem entailed by a local template specification, it is also a theorem of the system.

$$\eta_{s',s}(\varphi) \in Ax^{\vdash_{\Psi,s}} \text{ only if } (\varphi) \in Ax^{\vdash_{\Psi,s'}} \text{ whenever } s \leq s' \qquad (Spec)$$

The rule *Spec* reflects the specialization construct: each formula of the more general template specification carries over to the more specific template specification through the template specialization morphism $\eta_{s',s}$, and both are a theorem of the system specification.

$$\rho_i(\varphi) \in Ax^{\vdash_{\Psi,s}} \text{ only if } (\varphi) \in Ax^{\vdash_{\Psi,s_i}}, 1 \leq i \leq n, \text{ and } s = (s_1, ..., s_n) \qquad (Inc)$$

This rule reflects the incorporation construct: it says that the translations of formulas valid for a component is also valid in the aggregation.

$$(s \Rightarrow \varphi) \in Ax^{\vdash_{\Psi,s'}} \text{ only if } s \leq s' \text{ and } \eta_{s',s}(\varphi) \in Ax^{\vdash_{\Psi,s}} \qquad (SRefl)$$

This rule concerns reflection: if we can prove a formula in the specialization that only involves the vocabulary of the base specification, then this formula is valid for the base object if a specialization exists.

$$(\varphi) \in Ax^{\vdash_{\Psi,s_k+}} \text{ only if there is an aggregation } s = \mathbf{tuple}(s_1, ..., s_n) \qquad (IRefl)$$
$$\text{such that } s_k \in \{s_1, ..., s_k\} \text{ and the translation } \rho_k(\varphi) \text{ is}$$
$$\text{an element of } Ax^{\vdash_{\Psi,s}}$$

This rule state reflection for aggregations: if in the aggregation we can prove a formula that only involves the vocabulary of a component then that formula is a theorem of the component when considered in the aggregation.

The reflection rules *SRefl* and *IRefl* are very important since they define the impact an environment has on a local specification. When a component is embedded into an environment, the environment may impose further constraints on the component. Through the reflection rules, we may infer which constraints are added to the local context when a component is embedded into an environment. Thus, the resulting system theory is more than the sum of the component theories.

Definition 4.3.28 (*System Theory*)
Let $\Gamma = (\Psi, Ax)$ be a system specification. The *system theory* generated by Γ is the least set of formulas in $\mathcal{T}_\Psi$ satisfying the inference rules.   ∎

In [SSC92] the soundness of the global inference system has been shown.

4.3.4   Evaluation and Limitations

After having introduced the Object Specification Logic, let us now evaluate it as being the semantic framework behind the language TROLL.

First of all, let us give the reasons why we chose OSL as the framework to explain the semantics of TROLL:

- The way events and attributes are used very nicely reflects our intuition about events and attributes in TROLL. Especially the built-in predicates ▷ and ⊙ help in understanding the dynamics of an object system.

- The morphisms that model relationships between objects are simple but powerful, although we can only model static constructions.

- The logic provides us with an inherent notion of locality.

- The deductive system is reasonably simple, since only linear temporal logic is used. For modeling at the conceptual level, this is sufficient (although not satisfactory) since we model systems on a very high level of abstraction.

The logic has, however, a few drawbacks that should not be left unmentioned in the context of this thesis:

- The frame assumption made in OSL (inference rule *LocAttFrm* on page 74) is a very weak one that basically prevents spontaneous changes to occur. It does not implicitly prevent events from having more than the specified effects.

Here an *affects-relation* relating events and attributes on which those events may have an effect is a solution for a strict implicit frame rule.

- A notion of *fairness* (see [Fra86, MP92a]) is needed for the framework. As it is now, nothing can prevent the postponing of an enabled action until infinity.

- The framework only uses *future tense* temporal operators. In principle, the future tense fragment is sufficient for specifying characteristic behavior, but the modeling power decreases since e.g. straightforward safety constraints cannot be expressed directly.

- The morphisms between specifications are *static*. Thus, we can only model static relationships between objects directly in the logic. This may be considered sufficient for many applications but will cause problems when we are going to explain the semantics of TROLL in the logic.

- Nondeterminism is not well reflected in the models. The use of branching time models would be more adequate. A recent development makes use of an institutional framework to be independent of a particular temporal domain [SS93].

- The next-operator causes problems when composing object specifications. We have to consider local silent transitions that can be inserted whenever in the context of a global transition nothing occurs locally in a component.

Part II
Language Definition

5 Basic Constructs in TROLL

Before we are going to define the TROLL-language in detail let us present basic language constructs. Please recall that we use a slightly modified version of the language TROLL91 defined in [JSHS91].

In the next section, we will introduce and define the *base languages* underlying TROLL. These languages are employed in the sections of a TROLL specification to specify properties local to a specification unit. After that, we give an abstract overview of the higher-level modeling concepts in TROLL. Here we state the basic intuitions that are behind the structuring concepts.

5.1 Base Languages

This Section introduces the base languages that are used in TROLL specifications. These languages are used in the different sections of a TROLL template specification. These sub-languages are based on formal logical languages as introduced in Chapter 4. It is not a good idea, however, to give pure logical languages to the user. The base languages of TROLL provide a user interface to the pure logical languages. For the semantics, we will define a translation of the base language concepts into the corresponding concepts of OSL.

We will begin with a short introduction of language constructs for *data terms*. We will adapt language features known from algebraic specification (see Section 3.1) and will define a language for data terms.

Data terms are the building blocks of *temporal logic formulas*. In the second section of this chapter we will define a more readable temporal logic language and will define additional logical connectives that are defined in terms of the basic temporal operators. The basic as well as the additional temporal connectives can be used in specifications. Moreover, we will divide the resulting temporal logic language in a *future tense* fragment and a *past tense* fragment. This division is meaningful since only one of the fragments can be used in a section of a template specification.

The last section of this chapter describes the *pattern language*. Patterns define an ordering for a set of events, i.e. they define how certain events are ordered. In many cases (especially where we want to specify a rather restricted possible behavior) pattern language sentences are much more convenient that temporal logic specifications which become cumbersome and unreadable in such cases. The pattern language reminds more of process languages like CSP [Hoa85] which define an explicit ordering of events.

5.1.1   Data Language

Data values are used to quantify things and properties in our approach. We do not only need values but also expressions that denote values and we have to be able to evaluate such expressions. As we have seen in Section 3.1, the Abstract Data Types framework (or functional programming framework as it is called in [GM87]) is very useful for that purpose. Most approaches, however, use a somewhat clumsy notation. Moreover, TROLL specifications do not directly make use of algebraic specifications, they only include *data terms.*

Data terms are build over a data signature as defined in Definition 4.2.1 and a set of variables as defined in Definition 4.2.2. Please note that all elements of a data signature as well as the variables are elements of the *data alphabet.* Most of the elements of such a data alphabet are *user-definable,* only a few elements are considered to be predefined for TROLL.

In many approaches to abstract data types, only the prefix notation for functions is used. This is not the usual way functions are used by specifiers. They want also to use the infix or even the postfix notation. This is also allowed in languages of the OBJ-family [FGJM85]. Furthermore, function symbols may be equipped with priorities and attributes like in OBJ.

In the TROLL approach, *attributes* which describe properties of objects take values from data sorts. Thus, attributes are similar to variables in pure logical approaches. Some authors regard attributes to be functions that associate a data value with an instance. In OSL (which is underlying our approach), attributes are modeled as *predicates* that are interpreted functionally but may change dynamically over time.

In data terms within TROLL, attributes are used like variables, i.e. we are referring to the current value of the attribute.

Let $t_1, .., t_n$ be terms of sort $s_1, ..., s_n$ and $a(s_1, ..., s_n, s)$ an attribute symbol in *ATT*. Then

$$a(t_1, ..., t_n)$$

is an *attribute term* of sort s.

The semantics of data terms is described in Section 4.2. The only necessary adaption is that functions have to be transformed into prefix notation. Then the Definitions 4.2.4 and 4.2.6 apply without changes.

Formal Syntax

The formal syntax of data terms is given by the following rules:

<data_term>	::=	*<variable_id>*
		| *<constant_id>*
		| '('*<data_term>*')'
		| *<attribute_sel>*
		| *<prefix_fun_id>*'('*<arg_list>*')'
		| *<arg_list>* *<infix_fun_id>* *<arg_list>*
		| '('*<arg_list>*')'*<postfix_fun_id>*
<attribute_sel>	::=	*<attr_id>*
		| *<attr_id>*'('*<arg_list>*')'
<arg_list>	::=	*<arg_list>*','*<data_term>*
		| *<data_term>*

Semantics

The semantics of a data term is the value it denotes. Thus, term evaluation as defined in Definition 4.2.6 can directly be used. The only difference are attributes since in OSL they are treated as predicates. The denotation of an attribute term $a(t_1, ..., t_n)$ is defined as follows:

$$[a(t_1, ..., t_n)] = \begin{cases} [t] & \text{if } \models \, \triangleright \, a(t_1, ..., t_n, t) \\ \text{undefined} & \text{otherwise} \end{cases}$$

Thus, we have to interpret observations in OSL as data terms.

5.1.2 Temporal Logic Language

In this section we are describing the logical and temporal sub-languages of TROLL. We are assuming a *linear time* framework as defined for OSL (see Section 4.3) serving as a semantical framework. A good textbook on general temporal logic languages is [MP92a].

The first subsection is devoted to *state formulas*, i.e. sentences that specify the current state. These sentences are special cases of temporal languages, which are presented in the subsequent sections. In TROLL (as in TROLL91) we are providing two sub-languages:

- The *Future Tense Language* specifies characteristics of future states in life cycles. This language provides the full expressivity of the temporal fragment in OSL, although in a more user-friendly packaging.

- The *Past Tense Language* is used to specify preconditions the refer to states in the past. Thus, we are of the opinion that we do not need the full expressivity of past tense temporal logic here but can work with a simplified language. The simplification mainly prohibits nesting of temporal operators.

5.1.2.1   State Formulas

State formulas are sentences of which the truth value is evaluated over the current state only. State formulas have the usual first-order logic syntax. The basic building blocks of state formulas are *atomic formulas* over an alphabet V which consists of symbols of a signature and a set of variables.

- *Terms* over V are data terms as described in Section 5.1.1.

- A special term is an *event term*. Event terms are instantiated event symbols. Let $e(s_1, ..., s_n)$ be an event symbol in $EVT_{s_1,...,s_n}$ and $t_1, ..., t_n$ be data terms of sorts $s_1, ..., s_n$ respectively. Then

$$e(t_1, ..., t_n)$$

is an event term.

- *Atomic Formulas* are constructed over terms as follows:

  - If $t_1, ..., t_n$ are data terms and p is a data predicate, then $p(t_1, ..., t_n)$ (for prefix predicates) and $t_1\, p\, t_2$ (for infix predicates) are atomic formulas.

- A special atomic formula is a sentence

$$\mathbf{occurs}(e(t_1, ..., t_n))$$

where $e(t_1, ..., t_n)$ is an event term. It has the same meaning as the predicate $\odot$ in OSL and holds in a state where the event denoted by $e(t_1, ..., t_n)$ occurs.

From these atomic formulas, we may construct state formulas using established connectives known from first order logic.

- Each atomic formula is a state formula.

- If φ is a state formula, then **not** φ is also a state formula. It holds iff φ does not hold.

- If φ and ψ are state formulas, then φ **or** ψ is a state formula which holds iff φ or ψ hold.

- If φ and ψ are state formulas, then φ **and** ψ is a state formula; it holds iff both φ and ψ hold.

- If φ and ψ are state formulas, then φ **implies** ψ is a state formula; it holds iff φ implies ψ.

- If φ and ψ are state formulas, then φ **eq** ψ is a state formula.

- If φ is a state formula and x is a variable of type s, then (**forall** $x\colon s$)(φ) is a state formula.

- If φ is a state formula and x is a variable of type s, then (**exists** $x\colon s$)(φ) is a state formula.

Let us give a few examples for state formulas. We assume the signature for accounts (page 66) to be defined.

- The following formula states that in the current state an event of type `debit` occurs:

$$(\textbf{exists } \mathtt{m} : \mathtt{money})(\textbf{occurs}(\mathtt{debit(m)}));$$

- The next formula claims that a value of `Balance` that is less than 0 implies that the value of an attribute `Red` is equal to **true**:

$$(\mathtt{Balance} < 0) \textbf{ implies } \mathtt{Red} = \textbf{true};$$

- For savings accounts we may want to state that the balance must not be negative in a state:

$$\mathtt{Balance} >= 0;$$

Formal Syntax

The formal syntax of state formulas is given by the following production rules:

$<state_formula>$ ::= $<atomic_form>$
| **not** $<state_formula>$
| '('$<state_formula>$')'
| $<state_formula>$ **or** $<state_formula>$
| $<state_formula>$ **and** $<state_formula>$
| $<state_formula>$ **implies** $<state_formula>$
| $<state_formula>$ **eq** $<state_formula>$
| '('**forall** $<decl_list>$')' '('$<state_formula>$')'
| '('**exists** $<decl_list>$')' '('$<state_formula>$')'

$<atomic_form>$ ::= $<prefix_pred_id>$ '('$<data_term_list>$')'
| $<data_term>$ $<infix_pred_id>$ $<data_term>$
| **occurs**'('$<event_term>$')'

$<event_term>$ ::= $<simple_event_term>$
| $<selector><simple_event_term>$

$<simple_event_term>$ ::= $<event_id>$
| $<event_id>$'('$<data_type_list>$')'

Semantics

State formulas are translated into OSL formulas in a straightforward manner. Such a direct translation also serves for the definition of the semantics of state formulas in TROLL.

A note on the translation of atomic formulas is in order here: A possible observation of an attribute in TROLL is notated as

$$a(t_1, ..., t_n) = t$$

where $t_1, ..., t_n, t$ are data terms. This corresponds to a predicate

$$\triangleright\, a(t_1, ..., t_n, t)$$

in the Object Specification Logic OSL.

5.1.2.2 Future Tense Language

The future tense language specifies properties that must hold in future states. In TROLL, future tense sentences are used to specify dynamic constraints over attributes that have to be fulfilled in an admissible life cycle.

The language is developed from standard linear temporal logic such as used in OSL. An exception is the TROLL predicate

$$\textbf{initially } \varphi$$

for a state formula φ. A formula like that specifies a condition that holds in the state *after the birth* of an instance.

Consider the following future tense sentences as examples:

- To specify that the value of the attribute `Balance` is initialized to 0 after the birth of an account instance, the following formula is stated:

$$\textbf{initially } (\texttt{Balance} = 0);$$

- The next formula states that the balance of an account must always be greater or equal to 0 until the first deposit:

$$(\texttt{Balance} >= 0) \textbf{ until } ((\textbf{exists } \texttt{m})(\texttt{credit(m)}))$$

- Another example is the formula stating that the balance of a account must always be greater than the value of the attribute `CreditLimit`:

$$\textbf{always } (Balance >= CreditLimit)$$

Formal Syntax

The formal syntax is based on the syntax of template formulas (see Definition 4.3.3). We have a contextsensitive condition here saying that the keyword **initially** is not allowed to appear inside a future tense temporal formula.

<fut_formula>	::=	*<state_formula>*
		| **not** *<fut_formula>*
		| '('*<fut_formula>*')'
		| *<fut_formula>* **or** *<fut_formula>*
		| *<fut_formula>* **and** *<fut_formula>*
		| *<fut_formula>* **implies** *<fut_formula>*
		| *<fut_formula>* **eq** *<fut_formula>*
		| '('**forall** *<decl_list>*')' '('*<fut_formula>*')'
		| '('**exists** *<decl_list>*')' '('*<fut_formula>*')'
		| **initially** *<fut_formula>*
		| **next** *<fut_formula>*
		| **eventually** *<fut_formula>*
		| **always** *<fut_formula>*
		| *<fut_formula>* **until** *<fut_formula>*

Semantics

The semantics of other future tense sentences can be obtained by directly translating them into OSL formulas. Thus, we are not going to elaborate on that further at this point.

The only deviating concept in our future tense temporal language compared to OSL is the **initially** operator. A formula

$$\textbf{initially } \varphi$$

for a formula φ is translated into an OSL formula of the form

$$first \Rightarrow \bigcirc(\varphi)$$

where $first$ is a predicate that is only true in one particular state. Thus, we are working with an anchored temporal logic where an explicit initial state is given.

5.1.2.3 Past Tense Language

The past tense language is used to refer to the history of instances in *temporal preconditions* and permissions. Therefore, we do not need to employ full past tense temporal logic (see [MP92a, Section 3.3]) since this would put a too heavy burden on the specifier and would make a specification less readable and understandable for non-specialists. Difficulties with complex temporal preconditions have been reported by several (student) users of the language as well as by researchers that have listened to presentations of the TROLL91 language.

We restrict full past tense temporal logic (as used in the precursor language TROLL91) by *prohibiting nested temporal operators* This is not considered harmful since TROLL is very expressive and it has been realized that nobody used nested temporal operators so far in TROLL91.

Let us now introduce the various forms of past tense sentences informally. We assume that φ and ψ are state formulas as defined in Section 5.1.2.1.

- If φ is a state formula, then φ is also a past tense sentence.

- If φ is a state formula, then

$$\mathbf{previous}(\varphi)$$

  is a past tense sentence. The formula holds if the state formula φ was fulfilled in the precursor state provided we are not in the initial state.

- If φ is a state formula, then

$$\mathbf{once}(\varphi)$$

  is a past tense sentence. The formula is fulfilled if there exists a state before the current state in which φ holds.

- If φ is a state formula, then

$$\mathbf{has\ always}(\varphi)$$

  is a past tense sentence. The formula holds if the state formula holds in every valid previous state before the current state.

- If $e(t_1, ..., t_n)$ is an event term, then

$$\mathbf{has\ occurred}(e(t_1, ..., t_n))$$

  is a past tense sentence. This syntactical construct is an abbreviation for a frequently used longer sentence:

$$\mathbf{once}(\mathbf{occurs}(e(t_1, ..., t_n)))\mathbf{and\ not}(\mathbf{occurs}(e(t_1, ..., t_n)))$$

  saying that the event denoted by $e(t_1, ..., t_n)$ occurred sometime in the past and does not occur in the current state. While working with TROLL91, such a formula has been encountered very often in specifications; thus, we have introduced this abbreviation in TROLL.

- If φ and ψ are past tense sentences, then

$$(\varphi)\text{since last}(\psi)$$

  is a past tense sentence. Such a formula states that φ must be fulfilled after the last time ψ holds. Thus, we may also state conditions of which the validity is bounded towards the past.

- Past tense sentences may also be connected by the usual connective operators **and, or, implies**, and **eq**. The quantifiers **forall** and **exists** can also be used.

Formal Syntax

The following rules define the syntax of past tense sentences:

```
<past_formula>        ::=   <state_formula>
                        |  not <past_formula>
                        |  '('<past_formula>')'
                        |  <past_formula> or <past_formula>
                        |  <past_formula> and <past_formula>
                        |  <past_formula> implies <past_formula>
                        |  <past_formula> eq <past_formula>
                        |  '('forall <decl_list>')' '('<past_formula>')'
                        |  '('exists <decl_list>')' '('<past_formula>')'
                        |  once <state_formula>
                        |  has occurred'('<event_term>')'
                        |  has always <state_formula>
                        |  previous <state_formula>
                        |  <past_formula> since last <past_formula>
```

Semantics

The semantics of past tense sentences is defined by a translation of sentences into OSL. The difficulty is that OSL does not provide past tense temporal operators. Thus, past tense sentences must be represented by future tense sentences with reference to the initial state $k = 0$. It has been shown in [Pnu85] that every temporal logic formula containing past and future temporal operators is *initially equivalent* to a future tense temporal formula.

Here, it suffices to define how formulas of the forms

$$\mathbf{previous}(\varphi)$$

$$\mathbf{once}(\varphi)$$

$$\mathbf{has\ always}(\varphi)$$

and

$$(\varphi)\mathbf{since\ last}(\psi)$$

are translated since all other forms can be derived from them.

The translation is based on an additional predicate $P(s_1, ..., s_n)$ which is true in a state if the associated formula holds. The sorts $s_1, ..., s_n$ must correspond with the sorts of the free variables in the formula. Thus, a past tense sentence φ is expanded to

$$\varphi \Leftrightarrow P(x_1, ..., x_n)$$

where $x_1, ..., x_n$ are the free variables in φ. The predicate P can be regarded to represent historical information about the validity of past tense formulas when we cannot look into the past. The use of predicates to save historical information has been examined for TROLL91 specifications in [Sch92b, SS92].

1. $\mathbf{previous}(\varphi)$

   A formula of this form is translated into an OSL formula of the form

   $$\Box(\varphi \ \Rightarrow \ \bigcirc(P(x_1, ..., x_n)))$$

2. $\mathbf{once}(\varphi)$

   A formula of this form is translated into an OSL formula of the following form:

   $$\Box(\varphi \ \Rightarrow \ \Box(P(x_1, ..., x_n)))$$

3. $\mathbf{has\ always}(\varphi)$

   Such a formula is translated into

   $$\Box(\neg\varphi \ \Rightarrow \ \Box(\neg P(x_1, ..., x_n)))$$

4. $(\varphi)\mathbf{since\ last}(\psi)$

   Using additional predicates P_1, P_2 denoting the validity of the formulas φ and ψ, a sentence of the above form translates into

   $$(P_1(x_1, ..., x_n) \ \mathcal{W} \ P_2(y_1, ..., y_k)) \vee (\Box(P_2(y_1, ..., y_k) \ \Rightarrow$$
   $$\bigcirc(P_1(x_1, ..., x_n) \ \mathcal{W} \ P_2(y_1, ..., y_k))))$$

This schema states that P_1 is valid until the first time P_2 holds or P_1 holds and P_2 never holds, or if P_2 holds then afterwards P_1 holds unless the next time P_2 holds.

With these considerations we want to end the discussion of the past tense language in TROLL.

5.1.3 Pattern Language

The pattern language simplifies the description of restricted patterns of behavior over a set of events. Please note that in contrast to process specification languages like CSP [Hoa85] or CCS [Mil89] we in general do not describe the complete behavior of objects using this language, but we describe chunks of life cycles using the pattern language. The complete behavior can only be computed using the complete behavior specification. The description of such a deduction process would, however, be beyond the scope of this thesis. We will only describe the principal ideas behind various ways to specify behavior.

Our pattern language only provides means to specify *sequence* and *choice*. This is considered to be sufficient because

- parallelism should be inherent and not controlled by a specifier on a high level of abstraction

- synchronization should be specified by event calling (see Section 7.2)

- a loop is an operational concept that rarely occurs in conceptual modeling— thus, we decided not to support loops in TROLL.

Pattern specifications prescribes admitted and required ordering relationships between events.

The basic units of pattern specifications are

- *Event terms* as explained in Section 5.1.2.1 which denote events;

- *Cases* which specify an *external choice* between a number of (sub-)processes, and

- *Selection* which specify an *internal choice* between a number of (sub-)processes

Consider the following example: assume we want to model a part of the admissible behavior of an automatic teller machine (ATM). The signature includes the following event symbols:

- `issue_transaction(Amount:money [in])` (referring to the entering of an amount to be withdrawn from an account at an ATM);

- `send_TA(Bank:|Bank| [in],Acct:nat [in],Amount:money [in])` (designating the sending of the transaction to the bank that operates the ATM);

- `TA_OK` (denoting the successful commitment of the bank transaction);

- `TA_failed` (denoting the failure of a bank transaction) and

- `dispense_money(Amount:money [out])` (referring to the delivery of money).

We want to model that the event `issue_transaction(m)` is followed by an event `send_TA(B,n,m)` (where B and n are bound to appropriate values) which itself is followed by one of the events `TA_OK` or `TA_failed`. This selection is made by the environment. The event `TA_OK` should be followed by the event `dispense_money(m)`. This can be expressed by the following pattern specification:

```
ATM_TA(B,n) = issue_transaction(m)
                -> send_TA(B,n,m)
                    -> select
                            TA_OK -> dispense_money(m)
                        or  TA_failed
                        end select
```

Formal Syntax

Patterns can be generated using the following production rules:

<process_spec>	::=	*<process>*
			<process_id> **is** *<process>*
			<process_id>'('*<variable_list>*')' **is** *<process>*
<process>	::=	*<process>*'->'*<process_item>*
			<process_item>
<process_item>	::=	*<event_term>*
			<guarded_event>
			select *<process_enum>* **end select**
			case *<process_enum>* **esac**

$$<process_enum> \quad ::= \quad <process>$$
$$| \; <process_enum> \;\; \textbf{or} \;\; <process>$$

$$<guarded_event> \quad ::= \quad <application_cond><event_term>$$

Semantics

The semantics of process definitions will be given by a translation into formulas of the template language. Before we can do that, we have to *expand* the pattern specification. This is done by expanding process identifiers to the processes they name.

Basically, a pattern consists of a sequence of *blocks*. A block is either

- an *event term* $e_1(t_1, ..., t_n)$ or

- a *selection* **select**
$$e_1(t_1, ..., t_{k_1})$$
$$\textbf{or} \quad e_2(u_1, ..., u_{k_2})$$
$$\vdots$$
$$\textbf{or} \quad e_m(v_1, ..., v_{k_m})$$
end select

- or an *external choice* **case**
$$e_1(t_1, ..., t_{k_1})$$
$$\textbf{or} \quad e_2(u_1, ..., u_{k_2})$$
$$\vdots$$
$$\textbf{or} \quad e_m(v_1, ..., v_{k_m})$$
esac

The difference between external and internal choice cannot be described satisfactorily in OSL. A choice is external when the environment decides upon which event to take. A choice is internal when the instance (based on its internal state) by itself makes the choice between different alternatives. This difference cannot be modeled in our pattern language. Thus, in our semantical framework we may model external and internal choice likewise.

Moreover, an event term can be regarded as being a selection over just on alternative. Thus, an expanded pattern specification can be regarded as consisting of a tree of blocks of the following form:

case

$$P_1$$
$$\text{or} \quad P_2$$
$$\vdots$$
$$\text{or} \quad P_m$$

esac

where $P_1, ..., P_m$ are pattern specifications.

Since we are particularly interested in internal state changes concerning succeeding events (which may have effects lasting over a longer sequence), we may generalize further the pattern specification $P_1, ..., P_m$ into the starting blocks.

Thus, in general we are interested in changes in propositions concerning events in *subsequent states*. This is illustrated by Figure 5.1 where a tree starting in one event is depicted. The marked region identifies the region of interest: an event occurs and in the following state, we have the choice between several (here two) enabled events of which one and only one must occur. These events may be the starting points of sub-patterns. This makes again clear that a pattern specification includes *permissions and obligations*.

The basic unit of translation as explained before is a structure of the following form:

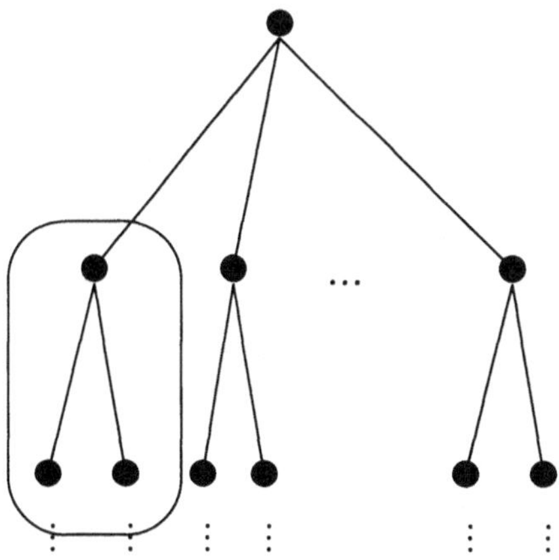

Figure 5.1: Pattern Structure

$$e(t_1, ..., t_n) \; \texttt{->} \quad \textbf{case}$$
$$e'_1(u_1, ..., u_{k_1})$$
$$\textbf{or} \quad e'_2(v_1, ..., v_{k_2})$$
$$\vdots$$
$$\textbf{or} \quad e'_m(w_1, ..., w_{k_m})$$
$$\textbf{esac}$$

Such a structure is translated into OSL formulas of the following kind:

$$\odot\, e(t_1, ..., t_n) \;\Rightarrow\; \bigcirc\,(\forall x_1, ..., x_n\colon \neg \rhd e(x_1, ..., x_n))$$
$$\wedge\, \bigcirc\,(\rhd e'_1(u_1, ..., u_{k_1}) \wedge ... \wedge \rhd e'_m(w_1, ..., w_{k_m}))$$
$$\wedge\, \bigcirc\left(\big|(\odot\, e'_1(u_1, ..., u_{k_1}), ..., \odot\, e'_m(w_1, ..., w_{k_m}))\right)$$

This formula states that when an event $e(t_1, ..., t_n)$ occurs, then no event of that kind is enabled in the next state, all events in the following choice of events are enabled and one and only one of those events actually occurs in the next state. The frame assumption (see Page 70) ensures that no currently disabled event is enabled in the next state unless explicitly stated.

A pattern can be transformed into a sequence of units as depicted in Figure 5.1. Thus, the translation of units have to be conjuncted. A problem arises when an event occurs several times in a pattern. We must therefore store the actual state in a predicate in such a case.

Another property of patterns is that they are required to occur before the death of an instance. Thus, we have to state that one of the starting events of a pattern must occur eventually.

Let us now turn to parameters. We explained above the intuitive statements for free variables in pattern specifications and parameters of pattern specifications. Recall that on the semantical level, we expanded all (nested) pattern identifiers. That means that we do not have to deal with pattern specification parameters on this level. A free variable in a pattern specification implies that it is *existentially quantified* in the entire pattern.

When we translate pattern specifications into sets of OSL formulas, we have to save the values of free variables for being used in states after their instantiation. This can be achieved by combining all (sub-)formulas as explained above into one formula where the free variables of event terms are bound by existential quantifiers. Consider the following pattern specification:

$$e_1(\texttt{n}) \; \texttt{->} \; e_2 \; \texttt{->} \; \textbf{select} \; e_3(\texttt{n}) \; \texttt{->} \; e_4$$
$$\textbf{or} \; e_5$$
$$\textbf{end select} \; \texttt{->} \; e_6$$

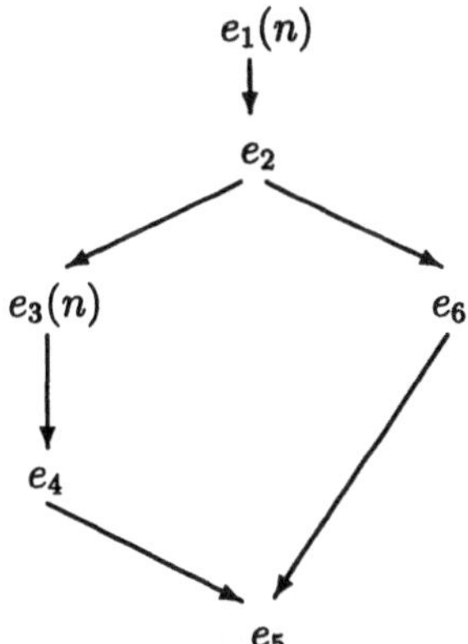

Figure 5.2: Pattern Example

The pattern is illustrated by Figure 5.2.

Please note that the parameter n of both event $e_1(n)$ and $e_3(n)$ is required to be bound to the same value at runtime. Note further that at least one instantiation of e_1 must occur. We now construct one formula using formula schemes as described above:

$$
\begin{array}{llll}
\exists n\colon [\\
\quad & \Diamond(\odot\, e_1(n)) & & (1) \\
\quad \wedge & \Box(\,\odot\, e_1(n) & \Rightarrow\ \bigcirc(\forall x\colon \neg\, \triangleright\, e_1(x)) & (2) \\
& & \wedge\, \bigcirc(\triangleright\, e_2) & (3) \\
& & \wedge\, \bigcirc(\odot\, e_2)\,) & (4) \\
\quad \wedge & \Box(\,\odot\, e_2 & \Rightarrow\ \bigcirc(\neg\, \triangleright\, e_2) & (5) \\
& & \wedge\, \bigcirc(\triangleright\, e_3(n) \vee \triangleright\, e_5) & (6) \\
& & \wedge\, \bigcirc(\odot\, e_3(n) \mid \odot\, e_5)\,) & (7) \\
\quad \wedge & \Box(\,\odot\, e_3(n) & \Rightarrow\ \bigcirc(\forall x\colon \neg\, \triangleright\, e_3(x)) & (8) \\
& & \wedge\, \bigcirc(\triangleright\, e_4) & (9) \\
& & \wedge\, \bigcirc(\odot\, e_6)\,) & (10) \\
\quad \wedge & \Box(\,\odot\, e_4 & \Rightarrow\ \bigcirc(\neg\, \triangleright\, e_4) & (11) \\
& & \wedge\, \bigcirc(\triangleright\, e_6) & (12) \\
& & \wedge\, \bigcirc(\odot\, e_6)\,) & (13) \\
\quad \wedge & \Box(\,\odot\, e_5 & \Rightarrow\ \bigcirc(\neg\, \triangleright\, e_5) & (14) \\
& & \wedge\, \bigcirc(\triangleright\, e_6) & (15) \\
& & \wedge\, \bigcirc(\odot\, e_6)\,) & (16) \\
]
\end{array}
$$

Thus, since we quantified the variable n existentially for the whole formula representing the pattern specification, it has to have the one single value throughout the whole pattern. In line (1) we state that the pattern must start eventually, i.e. that an instantiation of the first event must occur. Lines (2) to (4) state what is implied by the occurrence of $e_1(n)$, namely that no event of type e_1 is enabled in the next state (2), that the event e_2 is enabled in the next state (3) and that it must occur in the next state (4). In lines (5) to (7), the choice between $e_3(n)$ and e_5 is represented. Please note that n is quantified, thus it is set to a particular value. The following parts of the formula are constructed in a straightforward manner as explained above.

5.2 High-Level Language Concepts

In this Section, we are going to explain the high-level structuring concepts of TROLL in an intuitive way first in order to give an introduction to the language. The concepts are explained in more detail in the following chapters.

5.2.1 Templates

Templates are the basic structuring mechanism in TROLL. A template describes one *local aspect* of a conceptual object in a generic way. That is, no identifiers are associated with templates, it describes a prototype.

In a template, the local description of *structure* and *dynamics* of an aspect are integrated. A template consists of two parts:

- a *local signature declaration*, and

- a *local specification.*

A template is the description of an anonymous prototype object. An instance is structured as described by the template, its state is observable as defined in the template and the instance can only evolve along the lines specified in the template. An instance is an interpretation structure as defined in Definition 4.3.7.

In the local signature declaration, we declare the items that make up the interface of instances for the template. At the same time, the elements of the local signature make up the basic variable alphabet over which the specification is constructed. Furthermore, the local signature declaration can be seen as the specification of the *structure* of instances.

The local signature consists of the following classes of items:

- *Components* are specifications of component objects. These are specified separately.

- *Attributes* represent the observable properties of instances of a template. Attributes are typed, i.e. their values range over the carrier set of the associated sort.

- *Events* describe the local state transitions. They can have parameters which allow for the exchange of values in communications and for the description of effects of event occurrences on attribute values.

In the local specification, we describe the admissible observations over states and the admissible behavior of instances.

The admissible observations over states are specified using *constraints*. In TROLL, we may not only state static constraints which are invariant over the existence of an instance but also *dynamic constraints* which specify the possible evolution of observations over time. Dynamic constraints e.g. may require that the value of an observable property of an instance do not decrease or that there be a future point in time at which an observable property has a value greater than a constant.

The behavior of instances is described by the occurrence of events over time. In a template, we describe the admissible orderings of event occurrences. This makes up the possible behavior of instances. TROLL provides declarative and operational means to describe the behavior. The main characteristic of TROLL is that we do not have to describe the complete behavior patterns but we may restrict the possible behavior patterns to the admissible ones. This results in a more readable specification.

5.2.2 Object and Class Specifications

Based on a template specification, we may specify *single objects* in TROLL. An object is specified by associating an identifier with a template, thus *instantiating* the template.

A class specification defines *potential classes* of objects. A class specification defines a template for potential instances of the class and a *sort of identifiers* for instances of the class. Moreover, a class specification makes *key constraints* meaningful: no two instances in a class can have the same values for all key attributes.

A class in TROLL is regarded to be a collection of instances described by the same template and referenced through elements of a single data sort, the identifier sort. Thus, a class specification is not the primary modeling construct in TROLL.

5.2.3   Object Interaction

In TROLL, objects are encapsulated in the sense that only locally declared events
may alter the local attributes. In order to specify interactions between objects
that involve the specification of external effects on an instance's attributes, we
have to be able to *relate events* of (different) objects to specify that certain event
occurrences cause the occurrence of other events. In TROLL, interaction is always
synchronous, i.e. related events occur conceptionally simultaneously.

The basic mechanism to relate events is *event calling*. If an event e_1 calls another
event e_2, then an occurrence of e_1 implies an occurrence of e_2. If e_2 cannot occur
because it is not permitted to in the current state, the e_1 cannot occur, too.

5.2.4   Role Specifications

Roles specify temporary aspects of conceptual objects, e.g. the role `Employee` of
`Person`. The properties of the base aspect are valid for the role. A role can be
played several times in the life of a base object and several roles can be played
simultaneously. Therefore, the role concept is a dynamic concept in contrast to the
static concept of specialization.

A role specification introduces properties that extend the properties of the base
object. We may not override properties but may further restrict the behavior of
the base object when it plays a role.

5.2.5   Specialization

A specialization is a more detailed view on a conceptual object. A specialization
specification expands a base specification by additional properties. Again, we may
impose further restrictions on the base specification in the specialization.

Please note that we may not manipulate attributes of the base instance by events
local to the specialization directly. If an event local to the specialization has effects
on the base properties, then this must be specified using interaction between events.

The identifier sort of a specialization is a subsort of the identifier sort of the
base. That means that each identifier of a specialization instance identifies also a
base instance, but not vice versa.

5.2.6   Composite Objects

Composite objects have component objects which are related to the composite
object by the part-of relation. Composite objects in TROLL are associations: components in general are collections of objects.

- *Single components* are collections with at most one instance,

- *a set of components* is an ordinary collection, and

- *a list of components* is a multiset of components with an order on each element.

Component collections can vary over time—changes in the component collections are then described by events which are predefined for the composite object specification.

Composite object specification is a powerful modeling concept that allows to represent composite subsystems while taking in account also the behavior. Again, we have the restriction that local attributes may solely be altered by the occurrence of local events. Thus, effects of event occurrences that concern also other components or the composite object as a whole have to be described by interaction.

5.2.7 Relationships and Interfaces

In our approach, systems are constructed from components—thus, we must have a means to put local specification into a global context. On the language side, we should not be forced to specify context-sensitive properties in local specifications. Therefore, we introduced the concept of *relationship* into the language. A relationship describes interactions and dependencies between separately defined instances in terms of calling relations between events of the participating instances and in terms of global conditions.

A large system is unlikely to be specified as one unit composed from local specifications. Thus, we have to provide further levels of modularity. TROLL therefore includes also a concept of *interface*. An interface is a restriction of the local signature of an object specification. Thus, the semantics is give by those formulas in the theory of the underlying instance that only involve the restricted signature elements. Interfaces thus describe access restrictions to specifications.

Interfaces can be imported into templates and into system specifications. Furthermore, we may specify *system interfaces* which consist exclusively of interface definitions over the specification of an object system.

In the follwing three chapters, these high-level language features are defined.

6 Template, Object, and Class Specifications

This chapter describes the specification of basic syntactical units of system specifications in TROLL:

- *Templates* are descriptions of *prototypical objects*. A template does not describe one particular object but it describes the structure and dynamics of possible object instances.

- *Object Specifications* describe particular object instances. As such, they consist of a template and an *identifier* which serves also as the external name of the object instance.

- *Class Specifications* are descriptions of possible collections of object instances which are described by the same template. Furthermore, a class specification provides a set of possible identifiers for instances in this class and a possibility to access particular instances through key attributes.

It should be noted again that the TROLL approach is not class-oriented but object-oriented. The basic units of specification are *object descriptions* and not classes like in many other object-oriented approaches. Classes in TROLL are just collections of objects (which can be regarded on a semantical level as objects again).

In the following, we will start with presenting the language constructs which are used in template specifications.

6.1 Template Specification

The basic units of specifications in our approach are *templates*. Templates are descriptions of *prototypes* or generic instances. A template integrates the specification of the interface or signature of the prototype, the specification of the admissible observations and states, and the specification of the possible (or admissible) behavior

of instances. In the following subsections, we will describe the different sections of
a template specification and will define their semantics over the concepts described
in Chapter 3. In this chapter, we will restrict ourselves to the description of simple
non-structured templates that specify single instances.

A simple template specification consists of the following sections:

- Declarations of the local attribute symbols and types;

- Declarations of the local event symbols and parameters;

- Constraints on the admissible attribute values and on the evolution of attribute values over time;

- Effects of event occurrences on attribute values;

- Permitted event occurrences depending on the current state;

- Obligations to be fulfilled by instances of this template;

- Explicit process specifications, i.e. an explicit ordering of events.

In the following subsections, we will explain the parts of template specifications
in detail.

6.1.1   Local Signature Declaration

The signature of a template serves two purposes:

- it defines the interface of an object instance and

- it introduces the alphabet that can be used in a template specification.

According to the basic distinction between attributes or observable properties
and events in a local signature of a template we have to distinguish *attribute* symbols
and *event* symbols.

Attribute symbols allow to access that current values of attributes. Attribute
values can be observed and are by default not hidden. Attributes are *strongly typed*,
i.e. their values are elements of a sort that is the type of an attribute. These sorts
are assumed to be in the underlying universe of data types as explained before. The
types of attributes can be of arbitrary complexity, i.e. we could specify complex
types (e.g. **set**(...), see Definition 4.2.7) or identifier types. In particular, we are able
to use *user-defined* data sorts as attribute sorts which is of high practical usefulness
and has been an active research issue e.g. in databases [BB84, EN89, GH91]. In

contrast to [JSHS91] we do not require to explicitly import the data types to be used in the template.

The attribute section is marked by the keyword **attributes**. Let us start with an example of an attribute section. For a template that describes an account, we may want to specify the following attribute symbols:

```
attributes
   Number:  nat [constant key]
   Holder:  |Customer| [constant];
   Balance:  money; -- positive or negative
   Limit:  money;
```

The attribute **Number** takes natural numbers as values and represents the number of an account. This attribute does not change over time and serves as a key. The attribute **Holder** takes identifiers of type |Customer| and remains constant. The attributes **Balance** and **Limit** take values of type **money** and can change over time.

We have the following choice of characteristics of attributes:

- **key** for key attributes (see Section 4.1). There may not be two instances of the template in the same class having the same values for the key attributes. Keys are of relevance at the class level. On the level of instances, they serve as access paths for explicitly refer to an instance.

- **constant** for constant attributes. The values of these attributes are not allowed to change over time. This characteristic is only an abbreviation for a constraint requiring that the value never changes.

- **optional** for attributes that may not always have a defined value.

- **derived** for computed attributes. The values of derived attributes are computed from the values of other attributes.

The characteristics **derived** and **optional** must not be combined with any other characteristic.

The second class of items to be declared in the local signature declaration of a template are *event symbols*. Event symbols denote state transitions of instances. Event symbols may have parameters which are used to define effects of event occurrences or data flow during interactions. Event symbols are called *actions* in process specification, where an occurrence of an action is called an event.

In order to improve the readability of TROLL specifications, we are allowing event parameters to be named. The parameters are of a data sort. Furthermore, we may mark event parameters whether they are input parameters (**in**) or output parameters (**out**).

- *Input parameters* must be instantiated before the occurrence of an event.

- *Output parameters* are instantiated when the event occurs.

This distinction is not very important when templates are defined since in a template we are only specifying local properties. They are helpful, however, to give some hits towards the data flow in communications. As we will see in Section 7.2, we may have complex interactions and it may occur that data and control flows do not coincide. This has been examined also in [Saa93, Section 4.2.3].

We do not distinguish semantically between events that are suffered by the instance (i.e. they are called by another instance) and events that are *initiated* by the instance without external stimulus. In TROLL, both forms are possible. This aspect, as the data flow aspect, becomes relevant when we put objects into a system context. It is nevertheless helpful to be able to syntactically provide some guidance about the distinction since the language shall be suitable for requirements specification.

The event section is marked with the keyword **events**. Again we start with an example specification for a template for accounts:

```
events
    open(No:nat [in], Who:|Customer| [in]) [birth];
    close [death];
    debit(Amount:money [in]);
    credit(Amount:money [in]);
```

Here we have events denoting the opening of an account (the birth event **open**), the closing of an account (the death event `close`), and events that denote the withdrawal (`debit`) or deposit (`credit`) of amounts of money on the account.

We require at least one birth-event symbol to be declared. The occurrence of a birth event generates or creates an instance of the template. A death-event symbol is not required in a template—we may specify instances that are not supposed to disappear but (at least conceptually) to live eternally.

Event symbols may have the following characteristics:

- **birth** for birth-events.

- **death** for death-events.

- **active** for initiatives or active events.

An event marked by **active** can occur (if permitted) without being caused to occur by the environment. Here we may describe an object being capable of *initiating* the occurrence of events by itself. This is helpful when we describe e.g. actions of humans where the cause for their acting is irrelevant.

Formal Syntax

The syntax of a local signature declaration is defined by the following rules:

<attribute_section> ::= **attributes**
  <attr_decl_seq>`;`

<attr_decl> ::= *<att_symbol>*`:`*<sort_id>*[`[`*<attr_prop_list>*`]`]

<att_symbol> ::= *<attribute_id>*
  | *<attribute_id>*`(`*<sort_list>*`)`

<attr_prop> ::= **key**
  | **constant**
  | **derived**
  | **optional**

<event_section> ::= **events**
  <birth_event_seq>`;`
  [*<evt_decl_seq>*`;`]

<birth_event> ::= *<evt_symbol>*`[`**birth** [**active**]`]`

<evt_decl> ::= *<evt_symbol>*[`[`*<evt_property_list>*`]`]

<evt_symbol> ::= *<event_id>*
  | *<event_id>*`(`*<param_list>*`)`

<param> ::= *<param_id>*`:`*<sort_id>* `[`*<param_type>*`]`

<param_type> ::= **in** | **out**

<evt_property> ::= **death** | **active**

Semantics

Let us first define some very important substitutions that make the understanding of TROLL specifications much easier. TROLL puts emphasis on the specification of object dynamics—thus, we want to define the notions of the following classes of states using OSL:

- The predicate *prenatal* denotes situations where a particular object is not born yet. It is characterized by the fact that all the possible birth events are enabled but no other event is enabled and we may not observe any attribute.

- The predicate *alive* characterizes situations where the object is alive in the sense of TROLL: a birth event has occurred sometime in the past and no death event has occurred so far.

- The predicate *dead* characterizes states where a particular object is dead, i.e. its attributes cannot be observed anymore and no event is enabled.

In the following, we use the following abbreviations:

- $\vec{x}$ denotes a vector of substitutions for a vector of variables $x_1, ..., x_n$ of the proper sorts.

- $B \subseteq EVT$ is the set of birth events that can be extracted from a TROLL template.

- $D \subseteq EVT$ is the set of death events that can be extracted from a TROLL template.

Definition 6.1.1 (*Prenatal*)
The predicate *prenatal* is defined as follows:

$$
\begin{aligned}
prenatal \Leftrightarrow \quad & ((\forall b \in B \; \forall \vec{x}: \; \triangleright b(\vec{x})) \\
& \wedge (\forall e \in (EVT \backslash B) \; \forall \vec{x}: \; \neg \triangleright e(\vec{x})) \\
& \wedge (\forall a \in ATT \; \forall \vec{x}: \; \neg \triangleright a(\vec{x})) \\
& \wedge (\forall b \in B \; \forall \vec{x}: \; \neg \odot b(\vec{x})))
\end{aligned}
$$

∎

Definition 6.1.2 (*Birth*)
The predicate *birth* is defined as

$$
birth \Leftrightarrow ((\exists b \in B \; \exists \vec{x}: \; \odot b(\vec{x})))
$$

∎

Definition 6.1.3 (*Death*)
The predicate *death* is defined as follows:

$$death \Leftrightarrow (\exists d \in D \exists \vec{x} : \odot d(\vec{x}))$$

∎

Definition 6.1.4 (*Dead*)
The predicate *dead* is defined as

$$dead \Leftrightarrow ((\forall a \in ATT\ \forall \vec{x} : \neg \triangleright a(\vec{x})) \wedge (\forall e \in EVT\ \forall \vec{x} : \neg \triangleright e(\vec{x})))$$

∎

Thus, the predicate *dead* states that no local formula is valid. Thus, we may not deduce anything local to an object in states after the death of the object.

Definition 6.1.5 (*Alive*)
The predicate *alive* is defined as

$$alive \Leftrightarrow (\neg prenatal \wedge \neg dead)$$

∎

Now let us explain the translation of the declarations of local symbols in template signatures and axioms of a template specification.

In the **events**-section, we declare events to be **birth**-events, **death**-events, or **active** events. These declarations generate formulas in a template specification. In the following, we will show which general patterns are generated. Let $e_1(s_{1,1}, ..., s_{1,n_1}), ..., e_k(s_{k,1}, ..., s_{k,n_k})$ be event symbols and $x_{1,1}, ..., x_{k,n_k}$ be variables of the corresponding sorts.

1. **birth**

   If $e_1, ..., e_k$ are declared to be birth events of an object o, then after one of them occurred, no birth event may occur. Thus, a formula of the following form is generated:

   $$\left(\bigvee_{i=1,...,k} \odot e_i(x_{i,1}, ..., x_{i,n_i}) \right) \Rightarrow \Box \left(\bigwedge_{i=1,...,k} \neg \triangleright e_i(x'_{i,1}, ..., x'_{i,n_i}) \right)$$

2. **death**

   If $e_1, ..., e_k$ are specified to be death events, then after the occurrence of one
   of them only the predicate *dead* is valid. Thus, a formula of the following for
   is generated:

   $$\left(\bigvee_{i=1,...,k} \odot e_i(x_{i,1}, ..., x_{i,n_i}) \right) \Rightarrow \bigcirc(\Box(dead))$$

3. **active**

   The keyword **active** only becomes meaningful when objects are put into a
   system context. Active events may occur without being called.

In the **attributes**-section, we may mark attributes with the keywords **key**, **constant**,
derived, or **optional**. These markings generate formulas in a template specification
which constitute the semantics of the markings. Let an attribute $a(s_1, ..., s_n) : s$
be given.

1. **key**

   This marking establishes a functional relationship between the attribute and
   the identifier of an object. We model that (as mentioned in Section 3.5.2)
   as a state dependent predicate—which does not belong, however, to a local
   template and thus will not be considered in this section.

2. **constant**

   When we declare an attribute to be constant, it is not allowed to change
   during the lifetime of an object o. A declaration

   $$\mathsf{a}(s_1, ..., s_n) : s \; [\mathbf{constant}]$$

generates a formula of the form

$$\Box(\triangleright \mathsf{a}(x_1, ... x_n, x) \Rightarrow \bigcirc(\triangleright \mathsf{a}(x_1, ... x_n, x))) \; \mathcal{W} \; death$$

meaning that once an attribute value can be observed, this observation does
not change unless the object dies.

3. **derived**

   No formula is generated from the declaration of a derived attribute.

4. **optional**

   An attribute marked by **optional** must not always be observable. Thus, for all
   attributes that are not optional, we generate an axiom of the following form:

   $$\forall a \in ATT: \; alive \Rightarrow \exists \vec{x}: \; \triangleright a(\vec{x})$$

saying that when the object is alive then we must be able to observe the attribute.

Let us now go through an example to show the translation of a local signature declaration into a fragment of an OSL template specification.

Example 6.1.6 Let the following specification fragment be given:

```
-- template Account
  attributes
    No:  nat [constant key]
    Holder:  |Customer| [constant];
    Balance:  money; -- positive or negative
    Red:  bool [derived]
    Limit:  money;
  events
    open(No:nat [in], Who:|Customer| [in]) [birth];
    close [death];
    debit(Amount:money [in]);
    credit(Amount:money [in]);
    set_limit(Amount:money [in]);
  ...
```

The template signature generated from this specification text is the following:

$$\Theta = (IS, \alpha, ATT, EVT)$$

where

- $IS = \{|\text{Customer}|, |\text{Account}|\}$

- $\alpha = |\text{Account}|$

- $ATT = \left\{ \{\text{No}\}_{\rightarrow\text{nat}}, \{\text{Balance}, \text{Limit}\}_{\rightarrow\text{money}}, \{\text{Holder}\}_{\rightarrow|\text{Customer}|}, \{\text{Red}\}_{\rightarrow\text{bool}} \right\}$

- $EVT = \left\{ \{\text{open}\}_{\text{nat}\times|\text{Customer}|}, \{\text{close}\}, \{\text{debit}, \text{credit}, \text{set_limit}\}_{\text{money}} \right\}$

The specification fragment induces also the generation of a number of specification axioms that are included in the corresponding specification in the object specification logic. The following axioms over the signature Θ are generated from the specification fragment:

$$\odot \, \text{open}(n, c) \Rightarrow \bigcirc(\Box \neg \triangleright \text{open}(n', c')) \tag{1}$$
$$\odot \, \text{close} \Rightarrow \bigcirc(\Box dead) \tag{2}$$
$$\Box(\triangleright \text{No}(n) \Rightarrow \bigcirc(\triangleright \text{No}(n))) \, \mathcal{W} \, death \tag{3}$$
$$\Box(\triangleright \text{Holder}(c) \Rightarrow \bigcirc(\triangleright \text{Holder}(c))) \, \mathcal{W} \, death \tag{4}$$

6.1.2   Conditions in Specifications

In the specification of a template, *application conditions* can appear at several places. Here, we want to briefly discuss form and function of such conditions in template specifications.

Application conditions restrict the applicability of specification sentences in certain states. In general, application conditions may refer to the history of an instance. Thus, they are specified using past tense language as defined in Section 5.1.2.3.

It should be emphasized that application conditions are *necessary* conditions—a specification sentence is only valid in a state if the application condition holds in that state. Application conditions are specified as

> **only if**{ *<past tense sentence>* }

Application conditions may precede or follow the specification sentence they are connected to.

We will not present any examples for application conditions since there will be plenty of examples in connection with specification sentences in the following sections.

Formal Syntax

$$\langle application_cond \rangle \quad ::= \quad \textbf{only if} `\{ \; '\langle past_formula \rangle `\} \, '$$

Semantics

Application conditions basically are sentences of the past tense language of which the translation into OSL has been presented already in Section 5.1.2.3.

6.1.3   Derived Attributes

In templates, we may declare attributes of which the current values are *derived*, i.e. they depend on the values of other attributes. The reason to introduce language features for derived or computed attributes is that they may be used as abbreviations in specifications that improve readability and expressiveness.

In a TROLL specification, the rules for deriving attribute values from other values (called *derivation rules*) are given in the **derivation** section.

As an example consider the specification of the derived attribute Red in Example 6.1.6. We have two rules, each with an application condition. The first rule says that `Red` has the value **true** only if the value of the attribute `Balance` is less than zero; otherwise, it has the value **false**.

```
derivation
  only if{Balance < 0} Red = true;
  only if{Balance >= 0} Red = false;
```

Derived attributes can also have a value that is computed from the values of other attributes. Consider the following example:

```
derivation
  Maxwithdrawal = Balance + Limit;
```

The attribute `Maxwithdrawal` must be declared as being **derived**. In each state, the value of `Maxwithdrawal` is specified by the rule above.

Formal Syntax

$<$*derivation_section*$>$ ::= **derivation**
$<$*der_rule_seq*$>$'`;`'

$<$*der_rule*$>$ ::= $<$*application_cond*$><$*simple_derivation*$>$
| $<$*simple_derivation*$><$*application_cond*$>$
| $<$*simple_derivation*$>$

$<$*simple_derivation*$>$ ::= $<$*attribute_term*$>$ '`=`' $<$*data_term*$>$

Semantics

The semantics of simple derivation rules is rather straightforward. They can directly be translated into OSL. We model them as equivalences. Let $a(t_1, ..., t_n, t)$ an attribute term and d be a data term. Then formulas of the following scheme are generated:

$$(\triangleright\, a(t_1, ..., t_n, t)) \Leftrightarrow (d = t)$$

In the presence of an application condition c, the result of the translation is a formula of the following scheme:

$$[(\triangleright\, a(t_1, ..., t_n, t)) \Leftrightarrow (d = t)] \Rightarrow c$$

saying that the equivalence only holds if the condition c holds in the current state.

6.1.4   Constraints

In the **constraints**-section, we may specify constraints on attribute values (possible observations) as well as constraints on the evolution of attribute values over time, i.e. here we may also specify the admissible behavior of instances over time in a *declarative* way.

Declarative constraints as described here have been employed in data modeling for a long time. Traditionally, constraints are invariants, i.e. they have to be fulfilled in every state of the database. They have been generalized to invariants over the evolution of databases, i.e. they must be fulfilled in every admissible life cycle.

Constraints are specified using the future tense language defined in Section 5.1.2.2 since formulas that do not include any temporal operators are state formulas that can be handled in the same framework.

For our **Account** example, we may e.g. state the following constraints:

```
constraints
   initially (Balance = 0 and Limit = 0);
   eventually Balance > 100;
   Balance >= -Limit;
```

Constraints with the keyword **initially** state conditions to be fulfilled after the birth of an instance. Initial constraints can be used to initialize attributes of which the values are not supplied as parameters of a birth event. In our example above, the constraint

```
initially (Balance = 0 and Limit = 0);
```

states that the values of the attributes **Balance** and **Limit** are initialized as 0 after the birth of an instance.

A temporal sentence

```
eventually Balance > 100;
```

states that at least once in the lifetime of an **Account**-instance the value of the attribute **Balance** must be greater than 100.

The constraint

```
Balance >= -Limit;
```

is a traditional invariant—it has to be fulfilled in every state of an existing instance of **Account**.

Formal Syntax

$<$*constraint_section*$>$ ::=   **constraints**
$<$*fut_formula_seq*$>$';'

Semantics

As usual, the semantics of constraints are given as a translation into formulas of the template language.

Let φ be a sentence of the future tense language.

1. φ is a state formula
   A constraint of this form is translated into

   $$birth \Rightarrow (\Box(\varphi)) \; \mathcal{W} \; death$$

   meaning that the constraint must be fulfilled in each state of an existing object.

2. $\varphi \equiv$ **initially** p
   Such a constraint is valid in the state right after a birth event occurred. Thus, it is translated into
   $$birth \Rightarrow \bigcirc(p)$$

3. All other constraints can be translated only by renaming into formulas of OSL. Let p be a constraint. Then p is expanded to a formula of the scheme

   $$birth \Rightarrow (p) \; \mathcal{W} \; death$$

   claiming that the formula must be valid until the instance dies.

6.1.5 Effects of Event Occurrences

Recall that in our approach the state of an instance can only change due to the occurrence of an event. Thus, events induce state changes. Generally, an event occurrence changes the observations that can be made of the instance, i.e. it changes the values of attributes. Thus, we have to have a means to specify the *effects* of an event occurrence on attribute values.

To increase the readability of specifications, we do not use temporal logic for the specification of the effects of event occurrences on attributes but a special syntactic form. The general form of simple *effect rules* is the following:

> [*event term*] *attribute name* = *data term*

The intuitive meaning is that after the occurrence of an event denoted by *event term* the value of the attribute *attribute name* will be the value denoted by the *data term*. An important point is that the *data term* is evaluated in the state where the event occurs, i.e. the new value of the attribute *attribute name* is computed using the old state.

As an example consider the following rule which may be specified in the `Account` specification:

```
[debit(m)] Balance = Balance - m;
```

The intuitive meaning of this rule is that in the next state after the occurrence of the event denoted by `debit(m)` the value of the attribute `Balance` will be computed from the old value of `Balance` minus the value assigned to the actual parameter of the event `debit`.

Change rules may also be conditional. Then they have a form

> **only if**{ φ } [*event term*] *attribute name* = *data term*

where φ is a sentence of the past tense language. In this case, the rule may only be applied if the condition holds in the current state. As an example let us state the following rule for an `Account`:

```
only if{ m >= 0 } [set_limit(m)] Limit = m;
```

This means that only events of type `set_limit` having an actual parameter greater or equal to 0 have an effect on the value of the attribute Limit. Please note that we do not prohibit events `set_limit` with an actual parameter less than 0 to occur—they just do not have an effect on the attribute Limit.

As a last syntactical feature we may combine several effects of a single event occurrence in one change rule. This is illustrated by the following example:

```
[open(n,c)] No = n, Holder = c;
```

Here we actually have two rules with the same event term in brackets, we are thus allowing an abbreviation. This sentence can equally be stated as

```
[open(n,c)] No = n;
[open(n,c)] Holder = c;
```

Formal Syntax

<effects>	::=	**effects** *<chng_rule_seq>*`;`
<chng_rule>	::=	*<application_cond><simple_chng_rule>* | *<simple_chng_rule><application_cond>* | *<simple_chng_rule>*
<simple_chng_rule>	::=	`[`*<event_term>*`]` *<simple_derivation_list>*

Semantics

Intuitively, a change rule states which new attribute values can be observed after the occurrence of an event.

Please note that the implicit frame rule (inference rule *LocFrmAtt*) prevents attributes to change unless an event occurs (see page 73). This reflects the frame problem [HR92].

Thus, the translation is defined as follows. Let e be an n-ary event symbol (event term), a_1, a_2 be attribute symbols (attribute terms), and t be a data term of type s.

1. [e] $a = t$

   A simple change rule is translated into a formula

   $$\Box\,[\odot e \Rightarrow \bigcirc(\triangleright\, a(t))]$$

   meaning that in the state after the occurrence of e the observation $a(t)$ is possible.

2. **only if**$\{c\}$ [e] $a = t$

   A conditional change rule is translated into

   $$\Box\,[(\odot e \Rightarrow (\bigcirc(\triangleright\, a(t)))) \Rightarrow c']$$

   where c' is the object specification formula that is the translation of the past tense sentence c. The meaning is that the condition c' must be true in the current state otherwise the occurrence of the event e does not have any effects observable in the next state.

Effects of event occurrences on attribute values can be specified referring to the old values of the attributes. In the presence of such references, the old values have to be saved using variables. Thus, the translation of the change rule

```
[debit(m)]Balance=Balance-m;
```

is the following OSL formula:

$$\Box\,[(\odot\,\texttt{debit}(\texttt{m}) \wedge \rhd\,\texttt{Balance}(m')) \Rightarrow \bigcirc(\rhd\,\texttt{Balance}(\texttt{m}'-\texttt{m}))]$$

6.1.6 Permissions

Permissions concern the specification of the behavior of instances over time. Intuitively, permissions correspond to so-called safety rules and prevent something bad (i.e. an undesired behavior) to happen in that only permitted events may occur in a certain state. Permitted events, however, do not have to occur in a state. In contrast to constraints (which operate on attributes), permissions operate on events.

Permissions have the following general form:

> *event term* **only if**{ *enabling condition* }

The enabling condition must be fulfilled in the current state of the instance in order to enable the event denoted by *event term* to occur. Please note that permissions are only *necessary* conditions—enabled events are allowed to occur if the condition is fulfilled, but they do not have to occur.

Enabling conditions may refer to the history of an instance, i.e. to the (sequence of) states from the birth up to the current state. Thus, enabling conditions are specified in the past tense fragment of the temporal sublanguage defined in Section 5.1.2.3.

Simple permissions only refer to the attribute values in the current state. An example is

```
close only if{ Balance = 0 };
```

This permission states that the event **close** can only occur in the current state if the value of **Balance** is 0.

In permissions, we also may refer to the actual parameters of events. The following permission states that an event of type **debit** is only allowed to occur if its parameter has a value less than the value of the derived attribute **Maxwithdrawal**:

```
debit(m) only if{m <= Maxwithdrawal};
```

General permissions refer to the history of an instance. We can refer to properties that were valid in some previous state(s). Consider the following permission stating that the `Limit` can only be changed if a deposit of more than the new limit has been made since the last change in the `Limit`:

$$\texttt{set_limit(m) only if}\{\texttt{(exists m',m'')(m'>m and (has occurred(credit(m'))}$$
$$\texttt{since last occurs(set_limit(m'')))))}\}$$

Here, we have an example of a bounded permission—we only refer to the recent history after the last occurrence of the event `set_limit`.

Formal Syntax

<*permission_section*> ::= **permissions**
                                      <*permission_seq*>`;`

<*permission*> ::= <*application_cond*><*event_term*>
                              | <*event_term*><*application_cond*>

Semantics

The semantics of permissions in Troll-specifications are given as a translation into formulas of the template language.

In general, permissions have the form

$$e \text{ \textbf{only if}}\{p\}$$

where p is a sentence of the past tense language and e is an event term. This means that it must always hold that if p holds then e is enabled. In Section 5.1.2.3 we presented the representation of past tense sentences in terms of OSL. Recall that we used an additional predicate P denoting the holding of a past tense sentence (perhaps with free variables) in a state.

Let φ denote the representation of a past tense sentence p. Then a permission is translated into a formula of the following form:

$$\Box(\rhd e \ \Rightarrow \ (P(x_1,...,x_n) \wedge \varphi))$$

saying that it always must hold that e is enabled only if $P(x_1,...,x_n)$ holds in the current state.

6.1.7 Obligations

Obligations define long-term goals to be fulfilled by instances during their lifetime. Intuitively, instances can only die when their goals have been fulfilled. In the field of concurrent and distributed systems such goals are often called *liveness constraints*. Very often, an obligation specified is an obligation to die.

Obligations can be state-dependent, i.e. they only become an obligation if a state fulfills a precondition. Such preconditions (like preconditions for state changes) are specified using the past tense language. Thus, preconditions may refer to the history of the instance.

The simplest form of an obligation is an event term. This means that the events denoted by the event term must occur in the life cycle of an object. A simple example of this is the requirement that each account must be closed sometime, i.e. an instance must die:

```
close;
```

Please note that parameters of events are implicitly universally quantified if not otherwise stated. This is not always suitable. An obligation may have to be fulfilled for at least one element of a set. An example for this is that at least once a deposit must be made:

```
exists(m:money)(credit(m));
```

We also admit combinations of simple obligations. Obligations may be combined by disjunction and conjunction. For a disjunction, at least on of the alternatives must be fulfilled. As an example consider the obligation that at least one update (a withdrawal or a deposit) must occur in the life of an account:

```
exists(m:money)(credit(m) or debit(m));
```

For a conjunction, all the alternatives must be fulfilled. An example for such an obligation is the statement that both the deposit of money and the closing of the account are obligatory for accounts:

```
(exists(m:money)(credit(m))) and close;
```

Usually, obligations depend on the history of the object. As an example consider the following obligation stating that once some amount of money has been deposited on the account some amount (not necessarily the same) has to be withdrawn:

```
exists(m':money)(debit(m')) only if{exists(m:money)(has_occurred(credit(m)))}
```

A complete life cycle is only admissible if the obligations are fulfilled before the death event. Intuitively speaking, whenever the condition is true, the obligation is appended to a list of open goals to be fulfilled. When an obligation is fulfilled, it is removed from that list.

Formal Syntax

<obligations>	::=	**obligations** *<obl_expr_seq>*';'
<obl_expr>	::=	*<application_cond><obl_form>* | *<obl_form><application_cond>* | *<obl_form>*
<obl_form>	::=	*<obl_form>* **or** *<obl_form>* | *<obl_form>* **and** *<obl_form>* | '('*<obl_form>*')' | **exists**'('*<decl_list>*')''('*<obl_form>*')'

Semantics

Obligations correspond to what is called *liveness conditions* in concurrency theory and temporal logics [MP92a]. Thus, the formulas generated from obligations resemble such conditions.

The translation is defined as follows:

1. A simple obligation of the form

   $$e(\vec{x})$$

   is translated into the formula

   $$birth \Rightarrow \Diamond \odot e(\vec{x})$$

2. A conjuction or disjunction of obligations is translated into a conjunction and disjunction, respecitively.

3. An obligation of the form

   $$\mathbf{exists}(\vec{x})(e(\vec{x}))$$

is translated into a formula of the following form:

$$birth \Rightarrow \exists \vec{x}:\ \Diamond(\odot\, e(\vec{x}))$$

4. An obligation of the form

<obl_form> **only if**{ φ }

is translated into

$$\psi \Rightarrow (\varphi \wedge P(x_1, ..., x_n))$$

where ψ is the right hand side of the implication being the translation of <obl_form> and $(\varphi \wedge P(x_1, ..., x_n))$ is the translation of the application condition.

6.1.8   Commitments

Commitments are similar to obligations in that they also describe goals to be fulfilled by instances.

An object should try to fulfill commitments as soon as possible. In contrast to obligations (which just have to be fulfilled eventually), commitments describe short-term goals. Furthermore, if the environment of an object does not permit the object to fulfill its commitments, the specification is not violated.

The introduction of commitments into the language TROLL is motivated by its ability to be used in the early stages of system development. Here, we have to be able to state properties that an *operational* system shall fulfill, but do not want to have a fully operational semantics of the specification already in these early stages.

Commitments are usually conditional, i.e. they denote a causality for events that occur on the initiative of the object. Thus, commitments mean *sufficient* conditions that enforce the occurrence of events in the next state if these events are enabled. Commitments have been examined also in [DD92].

Formal Syntax

The formal syntax of commitments is very similar to the syntax of conditional obligations.

<commitm_section>   ::=   **commitments**
                                <comm_expr_seq>';'

<comm_expr>   ::=   <application_cond><obl_form>
                             | <obl_form><application_cond>

Semantics

Intuitively, commitments demand that specified events occur in the next state if they are enabled once the application condition holds in the current state. Thus, the application condition is *sufficient* for the occurrence of certain enabled events in the next state.

Let

$$e \ \textbf{only if}\{ \ c \ \}$$

be a commitment. Then a sentence of this form is translated into an OSL formula of the following form (φ is the translation of the past tense sentence c):

$$\varphi \wedge (P(x_1, ..., x_n) \Rightarrow \bigcirc(\triangleright e \Rightarrow \odot e))$$

6.1.9 Behavior Patterns

As explained already in Section 5.1.3, we find it desirable to have a more operational means to specify object behavior at our disposal. This is helpful when a declarative specification of very restricted behavior patterns becomes clumsy and lengthy.

The pattern language described in Section 5.1.3 is a language to specify parts of life cycles of instances. Please note again that behavior patterns do not necessarily prescribe all possible behaviors of an instance.

Example 6.1.7 Consider an *Automatic Teller Machine* (ATM). Let us give here only a fragment of the specification:

```
-- ATM template
  attributes
    No:nat [key];
    CashOnHand:money;
  events
    set_up [birth];
    remove [death];
    insert_card(AcctNo:nat [in]);
    input_PIN(PIN:nat [in]);
    check_w_bank(AcctNo:nat [in],PIN:nat [in]);
    card_OK; card_not_OK;
    input_Amount(Amount:money [in]);
    issue_TA(Acct:nat [in], Amount:money [in]);
    TA_ok; TA_failed;
    deliver_cash(Amount:money [in]);
```

```
      refill(Amount:money [in]);
  patterns
    insert_card(a) -> select
                           input_PIN(p) -> GO_ON(a,p)
                      or    cancel
                      end select;
    GO_ON(a:nat,p:nat) is check_w_bank(a,p) -> select
                                         card_OK -> DO_IT(a)
                                      or    card_not_OK
                                      end select;
    DO_IT(a:nat) is input_Amount(m)
                      -> issue_TA(a,m) ->select
                                       TA_ok -> deliver_cash(m)
                                    or    TA_failed
                                    end select;
```

An ATM here can only have a quite restricted behavior. When a card has been
inserted, we may proceed by entering the personal identification number (PIN)
or by a cancelation of the session (`cancel`). When the PIN has been entered,
the pattern GO_ON is followed saying that now an event representing the sending
of a message to the bank must occur (`check_w_bank`) which itself is followed by
one of the events `card_OK` or `card_not_OK` which represent the reply message by
the bank indicating whether the card is valid or not. If the card is valid, we
may go on following the DO_IT pattern which describes the process of issuing and
processing a bank transaction. We may enter the amount to withdraw, the a
message is sent to the bank (`issue_TA`) which is answered by one of the events
`TA_ok` or `TA_failed`. If the transaction finished sucessfully (`TA_ok`), the an event
representing the delivery of cash `deliver_cash` must occur. The variables a, p, and
m are bound to the same value throughout the pattern once they were initialized
in the events `insert_card(a)`, `input_PIN(p)`, and `input_Amount(m)`.

The events which are not part of the pattern specification can interleave at any
point of the life cycle provided they are enabled.

Formal Syntax

In the **patterns**-section, we may specify patterns using the syntax defined in Section
5.1.3.

<pattern_section>   ::= **patterns**
                            <process_spec_seq>`;`

Semantics

The semantics of behavior patterns is given by a translation into declarative temporal formulas in OSL. The translation was described already in Section 5.1.3. Let us give here only an example.

The $EVT_{|\text{ATM}|}$ component of the resulting template signature $\Theta_{|\text{ATM}|}$ includes the following elements:

$$\{\texttt{card_OK}, \texttt{card_not_OK}, \texttt{TA_ok}, \texttt{TA_failed}, \texttt{set_up}, \texttt{remove}\}_\epsilon$$
$$\{\texttt{insert_cardinput_PIN}\}_{\text{nat}}$$
$$\{\texttt{check_w_bank}\}_{\text{nat}\times\text{nat}}$$
$$\{\texttt{input_Amount}, \texttt{deliver_cash}, \texttt{refill}\}_{\text{money}}$$
$$\{\texttt{issue_TA}\}_{\text{nat}\times\text{money}}$$

The pattern specification is translated into the following OSL formulas:

$$
\begin{aligned}
\exists a, p, m: \big[\\
\Diamond(\odot\,\texttt{insert_card}(a)) \\
\wedge\ \Box(\ \odot\,\texttt{insert_card}(a) \quad &\Rightarrow\quad \bigcirc(\forall x: \neg\,\triangleright\,\texttt{insert_card}(x)) \\
&\quad \wedge \bigcirc((\forall x:\ \triangleright\,\texttt{input_PIN}(x)) \wedge \triangleright\,\texttt{cancel}) \\
&\quad \wedge \bigcirc((\exists x:\ \odot\,\texttt{input_PIN}(x))\ |\ \odot\,\texttt{cancel})\ ) \\
\wedge\ \Box(\ \odot\,\texttt{input_PIN}(p) \quad &\Rightarrow\quad \bigcirc((\forall x:\ \neg\,\triangleright\,\texttt{input_PIN}(x)) \wedge \neg\,\triangleright\,\texttt{cancel}) \\
&\quad \wedge \bigcirc(\triangleright\,\texttt{check_w_bank}(a,p)) \\
&\quad \wedge \bigcirc(\odot\,\texttt{check_w_bank}(a,p))\ ) \\
\wedge\ \Box(\ \odot\,\texttt{check_w_bank}(a,p) \quad &\Rightarrow\quad \bigcirc(\forall x,y:\ \neg\,\triangleright\,\texttt{check_w_bank}(x,y)) \\
&\quad \wedge \bigcirc(\triangleright\,\texttt{card_OK} \wedge \triangleright\,\texttt{card_not_OK}) \\
&\quad \wedge \bigcirc(\odot\,\texttt{card_OK}\ |\ \odot\,\texttt{card_not_OK})\ ) \\
\wedge\ \Box(\ \odot\,\texttt{card_OK} \quad &\Rightarrow\quad \bigcirc(\neg\,\triangleright\,\texttt{card_OK} \wedge \neg\,\triangleright\,\texttt{card_not_OK}) \\
&\quad \wedge \bigcirc(\forall x:\ \triangleright\,\texttt{input_amount}(x)) \\
&\quad \wedge \bigcirc(\exists x:\ \odot\,\texttt{input_amount}(x))\ ) \\
\wedge\ \Box(\ \odot\,\texttt{input_amount}(m) \quad &\Rightarrow\quad \bigcirc(\forall x:\ \neg\,\triangleright\,\texttt{input_amount}(x)) \\
&\quad \wedge \bigcirc(\triangleright\,\texttt{issue_TA}(a,m)) \\
&\quad \wedge \bigcirc(\odot\,\texttt{issue_TA}(a,m))\ ) \\
\wedge\ \Box(\ \odot\,\texttt{issue_TA}(a,m) \quad &\Rightarrow\quad \bigcirc(\forall x,y:\ \neg\,\triangleright\,\texttt{issue_TA}(x,y)) \\
&\quad \wedge \bigcirc(\triangleright\,\texttt{TA_OK} \wedge \triangleright\,\texttt{TA_failed}) \\
&\quad \wedge \bigcirc(\odot\,\texttt{TA_OK}\ |\ \odot\,\texttt{TA_failed})\ ) \\
\wedge\ \Box(\ \odot\,\texttt{TA_OK} \quad &\Rightarrow\quad \bigcirc(\neg\,\triangleright\,\texttt{TA_OK} \wedge \neg\,\triangleright\,\texttt{TA_failed}) \\
&\quad \wedge \bigcirc(\triangleright\,\texttt{deliver_cash}(m)) \\
&\quad \wedge \bigcirc(\odot\,\texttt{deliver_cash}(m))\ ) \\
\big]
\end{aligned}
$$

Basically, we specify that the event `insert_card` must occur sometime in the life of an ATM (i.e. the pattern must be started at least once), and the alternative ways to proceed when an event in the pattern occurs.

The length of the translation justifies again the additional language feature of pattern specifications.

6.1.10   Semantics of Simple Templates

In the previous sections, we have presented a systematic translation of TROLL language constructs into declarations and formulas in OSL. This translation gives us a template specification as defined in Definition 4.3.5. The semantics of a template specification in TROLL is the local theory generated from the corresponding template specification in OSL (see Definition 4.3.14).

Formal Syntax

$$
\begin{array}{lcl}
<\!template_spec\!> & ::= & [<\!attribute_section\!>] \\
 & & <\!event_section\!> \\
 & & [<\!derivation_section\!>] \\
 & & [<\!constraint_section\!>] \\
 & & [<\!effects\!>] \\
 & & [<\!permission_section\!>] \\
 & & [<\!obligation_section\!>] \\
 & & [<\!commitm_section\!>] \\
 & & [<\!pattern_section\!>]
\end{array}
$$

Semantics

Let us finally go through a complete template specification in TROLL and translate it into a template specification in OSL.

Example 6.1.8
Let us specify the template for `Account` objects as a whole:

```
-- Account template
  attributes
    No:  nat [key]
    Holder:  |Customer| [constant];
    Balance:  money; -- positive or negative
    Red:  bool [derived]
```

```
    Limit:  money;
  events
    open(No:nat [in], Who:|Customer| [in]) [birth];
    close [death];
    debit(Amount:  money [in]);
    credit(Amount:money [in]);
    set_limit(Amount:money [in]);
  derivation
    only if{Balance < 0} Red = true;
    only if{Balance >= 0} Red = false;
  constraints
    initially (Balance = 0 and Limit = 0);
    eventually Balance > 100;
    Balance >= -Limit;
  effects
      [open(n,c)] No = n, Holder = c;
    only if{ m >= 0 } [set_limit(m)] Limit = m;
      [debit(m)] Balance = Balance - m;
      [credit(m)] Balance = Balance + m;
  permissions
    only if{ (Balance-m) >= -Limit } debit(m);
    only if{ Balance = 0 } close;
  -- end Account template
```

The local specification in OSL generated from this TROLL specification looks as
follows. The following template signature is generated from the TROLL template:

$$\Theta = (IS, \alpha, ATT, EVT)$$

where

- $IS = \{|\texttt{Customer}|, |\texttt{Account}|\}$

- $\alpha = |\texttt{Account}|$

- $ATT = \Big\{\{\texttt{No}\}_{\to\texttt{nat}}, \{\texttt{Balance}, \texttt{Limit}\}_{\to\texttt{money}}, \{\texttt{Holder}\}_{\to|\texttt{Customer}|}, \{\texttt{Red}\}_{\to\texttt{bool}}\Big\}$

- $EVT = \Big\{\{\texttt{open}\}_{\texttt{nat}\times|\texttt{Customer}|}, \{\texttt{close}\}, \{\texttt{debit}, \texttt{credit}, \texttt{set_limit}\}_{\texttt{money}}\Big\}$

The set of axioms includes the following formulas:

$$Ax_{\texttt{Account}} = \{ \quad \odot\,\texttt{open}(n,c) \Rightarrow \bigcirc(\square\neg \rhd \texttt{open}(n',c')),$$
$$\odot\,\texttt{close} \Rightarrow \bigcirc(\square \mathit{dead}),$$

$$\Box(\triangleright \mathtt{No}(n) \Rightarrow \bigcirc(\triangleright \mathtt{No}(n))) \; \mathcal{W} \; death,$$
$$\Box(\triangleright \mathtt{Holder}(c) \Rightarrow \bigcirc(\triangleright \mathtt{Holder}(c))) \; \mathcal{W} \; death,$$
$$birth \Rightarrow \bigcirc(\triangleright \mathtt{Balance}(0)),$$
$$birth \Rightarrow \bigcirc(\triangleright \mathtt{Limit}(0)),$$
$$birth \Rightarrow \Diamond(\triangleright \mathtt{Balance}(m) \wedge (m > 0)) \; \mathcal{W} \; death,$$

Let us briefly illustrate an interpretation structure for the template above. In Figure 6.1 we have associated with points in time (natrural numbers) sets of valid predicates. Due to limited space, we use the conventions that lower case variables $(n, c, ...)$ are universally quantified variables if not otherwise stated and upper case letters $(N, C, ...)$ denote values.

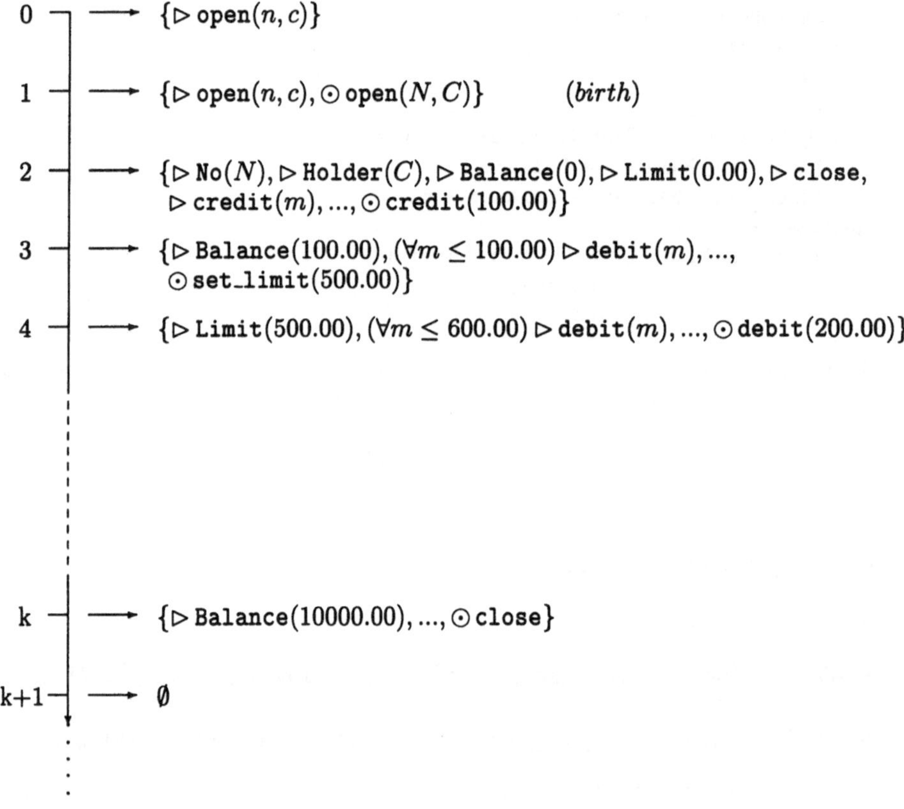

Figure 6.1: A Life Cycle of an Instance

6.2 Templates versus Classes

In the last Section we presented the concept of template. On first sight, templates as prototypes of object instances could be regarded as the type of the object. A template, however, is not enough to define a *class* of objects. This is in contrast to the opinion of a number of authors, especially from the field of object-oriented programming. They are of the opinion that a class is just a template. They neglect the problem of object identifiers. For them, those identifiers are provided by the system, they are *untyped.*

For conceptual modeling, however, this is not always suitable since we need an implementation-independent notion of identifiers. Furthermore, for a formalism dealing also with other relationships than references to other objects that make access possible it is not sufficient to deal with untyped identifiers. In object-oriented programming, access to objects is possible through references that are usually provided as the values of object methods.

Thus, just the template is not enough to define a class. We need at least a data type for the identifiers. The issue of identifiers has been discussed already in Section 3.5.2. Providing a data type for identifiers of possible instances of a class yields a carrier set of possible values for identifiers. We call this carrier set an *identification space.*

A class specification defines the *possible extensions* of the class—it is intensional. The current extension of a class is a set of *instances.* An instance is a mapping of an identifier to the current state of a model of the associated template of the class specification.

Only in the light of a class specification, keys and the induced constraints become meaningful. Keys are used to access instances of collections. Therefore, the values of key attributes must be distinct for all existing instances of a class at any moment in time.

6.3 Class Specification

Classes in TROLL are regarded to be *collections of objects* that are described by a common template. A *class specification* describes the potential extensions of a class.

A class specification adds the following to a template specification:

- Associated with a class specification is a *carrier set* for the identifier sort. Thus, we have at our disposal a set of identifiers that identify the *possible instances* of the class. This set is called the *identifier space* in TROLL.

- Key attributes are meaningful only in collections, especially in classes. Over a collection, *key constraints* are valid stating that there currently do not exist two instances having the same values of the key attributes.

It is not always desirable to have completely abstract identifier sorts. In many cases, identifier sorts are constructed over data domains as explained in Section 3.5.2. Thus, in TROLL we may (similar to the approach in Glider [DDRW91] or to TROLL91 [JSHS91]) declare the identifier sort to be constructed. Basically, this is achieved by defining constructor functions for the abstract data type that serves as the identifier space for a class. Consider the following example where we declare the identifier space for the class **Account** to be constructed over the atomic data type **nat**:

Example 6.3.1 Let the following specification fragment be given:

```
class Account over nat

   ...
```

Implicitly, this implies that the (internal) specification of the data type providing the identifier space includes the following operation definition:

```
data type |Account|
   sorts nat, |Account|;
   opns
      Account:nat  →  |Account|;
   ...
```

Thus, we may use the (data type) operation `Account(nat)` to generate an identifier (i.e. a data value) of sort `|Account|`.   ∎

This concept becomes important when it comes to referencing objects through their identifier. This will be considered in the next chapter in Section 7.1.

A special kind of class is a *single object*. In TROLL we allow the specification of single objects. A single object is a class with one and only one possible instance. The identifier space is restricted to consist of just one element, i.e. we have one constant in the associated data type signature which is the name of the object.

Example 6.3.2 An example for the specification of a single object is the following specification. We want to model a single bank object **Bank**:

```
object Bank
   attributes

      . . .

end object Bank;
```

The identifier space would be defined (in some algebraic specification language) as follows:

```
data type  |Bank|
  sorts  |Bank| ;
  opns
    Bank:  →  |Bank| ;
end data type  |Bank| ;
```

The constant of the identifier space definition is also the external name of the only possible instance—there is no need to define key attributes.   ■

Formal Syntax

The syntax of class and object specifications is given through the following rules:

<class_spec>   ::=   **object** *<class_id>*
                                      <template_spec>
                              end object *<class_id>*‘;’
                              | **class** *<class_id>* [**over** *<data_sort_list>*]
                                      <template_spec>
                              end class *<class_id>*‘;’

Semantics of Class Specifications

The semantics of a class specification is a two-level construction in OSL. The first step is the aggregation of all possible instances into an aggregation using incorporation morphisms. This aggregation is called *class base* in our approach. Without loss of generality, we claim that the cardinality of identifier spaces be countable. The second step is a specialization over this class base in order to be able to specify additional properties. Thus, we have incorporations $\rho_1, \rho_2, \ldots$ of the template specification TemplateSpec defining a generic instance into the ClassBase. The template specification ClassBase itself is specialized into the template specification Class because only then we are able to declare additional symbols and to state additional axioms. This construction is illustrated by Figure 6.2.

In the following, we denote by |Class| the sort of identifiers for the *instances* of the class Class, i.e. the local sort of the TemplateSpec. The additional symbols which are generated for a class specification are the following:

- An attribute **Extension** of type **set(**|Class|**)** represents the set of identifiers of currently existing instances of the class Class. The value of this attribute can be derived from the state of the aggregation ClassBase (see below).

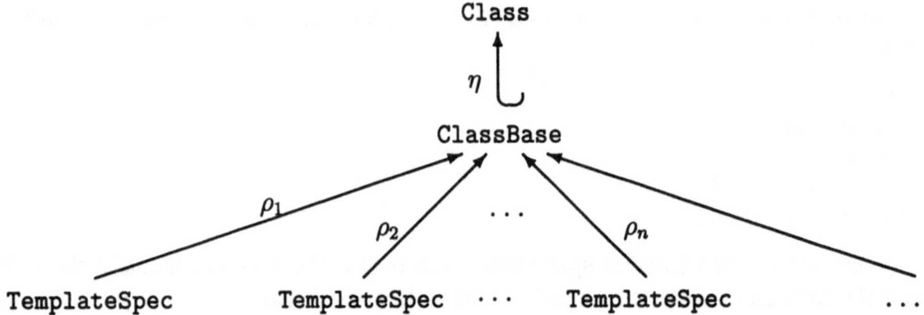

Figure 6.2: Two-Level Construction of Class Specification Semantics

- Let $s_1, ..., s_k$ be the sorts of the key attributes of the template of **Class**. An attribute $\textbf{KeyMap}(s_1,...,s_k,|\textbf{Class}|)$ denotes the current mapping of key attribute values to identifiers.

The following axioms are generated for the representation of a class. **class** is assumed to be declared as a variable over the local identifier sort for the template specification **Class**.

- A *key constraint* states that there may not be two distinct instances in one state that have the same values of the key attributes. This results in a formula of the following scheme:

$$(\textbf{class.} \triangleright \textbf{KeyMap}(k_1, ..., k_n, \pi_i) \wedge \textbf{class.} \triangleright \textbf{KeyMap}(k_1, ..., k_n, \pi_j)) \Rightarrow (i = j)$$

- Let $key_1, ..., key_n$ be key attributes of an instance template. The specification of the current value of the attribute **KeyMap** is derived using a formula of the following scheme:

$$(\textbf{class.} \triangleright \rho_i(key_1)(x_1) \wedge ...\wedge$$
$$\textbf{class.} \triangleright \rho_i(key_n)(x_n)) \Rightarrow \textbf{class.} \triangleright \textbf{Keymap}(x_1, ..., x_n, \pi_i)$$

The formula states that if we may observe in an incorporated component object π_i the values $x_1, ..., x_n$ of the key attributes $key_1, ..., key_n$, then we may observe the **KeyMap** attribute denoting the relationship between the key attributes and the identifier of the component. Please recall that a morphism ρ_i denotes the incorporation of the $i-$th component template into the composite object.

- In a similar way we specify the derivation of the current value of the attribute `Extension`:

$$(birth(\pi_{i_1})$$
$$\wedge birth(\pi_{i_2})$$
$$\vdots$$
$$\wedge birth(\pi_{i_n})$$
$$\wedge \texttt{class.} \rhd Extension(C)) \;\Rightarrow\; \bigcirc(\texttt{class.} \rhd \texttt{Extension}(C')$$

  where $C' = C \cup \{\pi_{i_1}, \pi_{i_2}, ..., \pi_{i_n}\}$ and $i_1, i_2, ..., i_n$ are distinct. This formula states that iff components $\pi_{i_1}, \pi_{i_2}, ..., \pi_{i_n}$ of the class base are born and we may observe the set C to be the current extension of `class`, then the identifiers $\pi_{i_1}, \pi_{i_2}, ..., \pi_{i_n}$ are in the `Extension` in the next state. Please note that there may be several instances that are born simultaneously.

Finally, let us illustrate the semantics of a class using our Example 6.3.1 from above.

Example 6.3.3 The signature $\Theta_{(\texttt{Account}_{i_1},...,\texttt{Account}_{i_n})}$ is the signature of the aggregation `AccountClassBase`. We have injection morphisms

$$\rho_i \colon \Theta_{\texttt{Account}} \to \Theta_{(\texttt{Account}_{i_1},...,\texttt{Account}_{i_n})}, i \in |\texttt{Account}|$$

incorporating the signature of each possible instance into the `ClassBase`. The class signature $\Theta_{\texttt{AccountClass}}$ is obtained by an inclusion morphism

$$\Theta_{(\texttt{Account}_1,...,\texttt{Account}_n)} \hookrightarrow \Theta_{\texttt{AccountClass}}$$

of the `AccountClassBase` into the `AccountClass`. The additional signature elements of $\Theta_{\texttt{AccountClass}}$ are the following:

- $\texttt{Extension}_{\mathbf{set}(|\texttt{Account}|)} \in ATT_{\texttt{AccountClass}}$

- $\texttt{KeyMap}_{nat,|\texttt{Account}|} \in ATT_{\texttt{AccountClass}}$

The set of axioms $Ax_{\texttt{AccountClass}}$ includes a translation of the axioms of each instance. Let φ be an axiom in the template specification $\mathbf{T}_{\texttt{Account}}$. Then the translation $\rho_i(\varphi)$ is an axiom in $Ax_{\texttt{AccountClass}}$ which represents a property of the $i-th$ instance. The additional axioms in $Ax_{\texttt{AccountClass}}$ are the following:

- The *key constraint* saying that whenever we find identical values of the attribute `No` in instances, we are talking about at most one instance.

$$ac. \rhd \rho_i(\texttt{No}(n)) \wedge ac. \rhd \rho_j(\texttt{No}(n)) \Rightarrow i = j$$

- The derivation of the value of the attribute **Extension**:

$$ac. \odot \rho_i(\mathbf{open}(n, h)) \wedge ac. \triangleright \mathbf{Extension}(C) \Rightarrow \bigcirc(ac. \triangleright \mathbf{Extension}(C \cup \pi_i))$$

- The derivation of the attribute **KeyMap**:

$$ac. \triangleright \rho_i(\mathbf{No}(n)) \Leftrightarrow ac. \triangleright \mathbf{KeyMap}(n, \pi_i)$$

$\blacksquare$

These considerations conclude the chapter describing the basic structuring mechanisms in TROLL. In the following two chapters, we are going to describe further structuring mechanisms for structured conceptual objects and the specification of systems composed from objects.

7 Relating Specifications

In this Chapter we present concepts and language features to describe closely related specifications. Before we are going to present the concepts of *specialization*, *roles*, *abstractions*, and *composite objects*, we will elaborate on two topics which provide basic mechanisms in complex specifications: *object referencing in collections* and *interactions between components*.

7.1 Referencing in Collections

After having described the specification of classes as collections, we are going to elaborate again on object referencing, here on the syntactical level. The topic of referencing objects in a certain context is not only a syntactical problem, however— it is a central problem of specifying composite dynamic systems.

This problem has been addressed also in [KS92] and in the report about the TROLL91 language (being a predecessor of the language described in this thesis) [JSHS91]. The solution in both cases was that objects can only be *accessed* in a local context where the objects to be referenced are components of the object in which the referencing should occur. Thus, pure references like used as special attributes are of no use since the possibility of referencing does not imply the possibility of accessing any property of the referenced instance.

Another problem is raised by the fact that we allow for state-dependent keys and referencing through key values. If an object is not alive yet (i.e. no birth event has occurred so far), we have no key map available but want to access the birth event of a particular instance. In TROLL91, this problem could not occur since the identification of an instance was also used for referencing the instance and the identifier remained constant forever (it was even there before the instance was born). In Section 3.5.2 we have discussed the drawbacks of that approach and have argued for another solution.

We solve the problem by overloading symbols. Let **CName** be the name of a class and assume it to be defined over a list of data sorts $s_1, ... s_k$. Then the identifier data type provides a constructor operation $\texttt{CName} : s_1 \times \ldots \times s_k \rightarrow |\texttt{CName}|$. This makes the generation of an identifier of type $|\texttt{CName}|$ by providing a vector of data items $\vec{x}$ possible.

When an instance is alive, we have at our disposal a key map that relates key attribute values and identifiers. The following function (called *Selector* which itself is named after the collection, thus this name is overloaded) yields an identifier:

$$\texttt{CName}(\vec{x}) = \begin{cases} \texttt{CName}(\vec{x}) & \text{if } \texttt{CName} \in \Omega_{|\texttt{CName}|} \text{ and} \\ & \forall \vec{x}, o: \neg \triangleright \texttt{KeyMap}(\vec{x}, o) \\ o & \text{if } \triangleright \texttt{KeyMap}(\vec{x}, o) \end{cases}$$

In connection with a TROLL template specification (which states that in prenatal state only a birth event is enabled) this definition implies the following:

- Only the birth event can be invoked using a generator function which is defined for the identifier type (which is a *data type*).

- When an instance is alive, it can only be referenced through the key attribute values (if key attributes are defined in the associated template).

Example 7.1.1 Consider the specification of the class $\texttt{Account}$ (see Example 6.3.1). In Example 6.3.1 we have defined the identification space for the class $\texttt{Account}$, including a function $\texttt{Account}:\texttt{nat} \rightarrow |\texttt{Account}|$ mapping a natural number to an identifier. If the object is not born yet, the operation

$$\texttt{Account}(1234)$$

yields a value of sort $|\texttt{Account}|$, say a. The event $\texttt{open}$ of $\texttt{Account}$ was declared to have parameters of sort $\texttt{nat}$ and $|\texttt{Customer}|$. Assume that the instance identified by a has not been born yet. Then the birth could be described by e.g.

$$\texttt{Account}(1234).\texttt{open}(45678, c)$$

where c is a value of sort $|\texttt{Customer}|$. After the occurrence of this event, the term

$$\texttt{Account}(45678)$$

yields the value a from above!   ∎

Selectors are used frequently in TROLL specification when interaction and dependencies between components are described.

Formal Syntax

Selectors can prefix attribute terms and event terms. The syntax of such prefixes is the following:

$$<selector> \quad ::= \quad <selector><select_id>`.`$$
$$| \; <select_id>`.`$$

$$<select_id> \quad ::= \quad <sel>[`(`<data_term_list>`)`]$$

$$<sel_id> \quad ::= \quad <class_id>$$
$$| \; <component_id>$$

7.2 Interaction Specification

A key concept for the language constructs described in this chapter is the specification of *interactions*. This is due to the *encapsulation principle*: *Attribute values can solely be changed by locally defined events.* Thus, we may not specify direct effects of other events on attributes. The encapsulation principle implies that the only way for objects to have an impact on the observations and the behavior of other objects is by relating events of both objects.

We may only specify interactions between locally known event symbols—it is not possible in a specification to refer to events of objects that are not related to the local object.

The basic concept to relate events of two or more objects is *event calling*. Intuitively, if an event e_1 calls another event e_2, then the occurrence of e_1 implies the *synchronous* occurrence of the event e_2. Thus, event calling means asymmetric synchronous interaction.

In a specification, event calling is written as

$$e_1 \; \text{>>} \; e_2;$$

if an event e_1 calls an event e_2.

Calling may also be conditional: A calling relation only is applicable if the condition holds in the current state. A condition may refer to the history, i.e. it is a sentence of the past tense language. Conditional calling is notated as

$$(e_1 \; \text{>>} \; e_2) \; \textbf{only if} \; \{condition\};$$

We may also specify *multicasting*, i.e. the calling of a specified set of other events. If an event e calls a set of events $e_1, ..., e_n$, this is notated as

$$e \gg e_1, e_2, \ldots, e_n;$$

As an abbreviation, we may use the notation

$$e_1 \; \texttt{==} \; e_2;$$

for calling in both directions, i.e. these events always occur simultaneously (*event sharing*).

The calling-concept in TROLL supports *synchronous interaction* only. In this version, calling is *strict*: a calling event can only occur if the called event can occur (thus, it must be enabled)!

Furthermore, the calling concept is (although it has operational flavor) *descriptive*: an event may call other events that itself may call events. Thus, we may not determine the set of synchronously occurring events at specification time since this set is state-dependent.

Using parameters, we may specify the exchange of values during interactions (*communication*). This exchange of values is specified through variables. The *flow of information* is determined by instantiated variables. The *unification* of uninstantiated variables with instantiated ones constitutes the flow of information.

Example 7.2.1 Assume that we declared the following events:

```
issue_Transaction(AcctNo:nat, Amount:money);
process_TA(No:nat [in], Amt:money [in]);
```

Now let us specify the calling of `process_TA` by the event `issue_Transaction` with the exchange of information through the parameters assuming that the parameters of `issue_Transaction` are instantiated:

```
issue_Transaction(a,m) >> process_TA(a,m);
```

For the syntactical characterization of the information flow we may use the markings **in** and **out** for event parameters (see page 111). Parameters marked by **in** are instantiated from outside the object during an interaction whereas parameters marked by **out** have to be instantiated with a value when the interaction takes place. Consequently, we may have information flows in both directions during an interaction. This is illustrated by the following example:

Example 7.2.2 Assume that we have specified events that model the request for the current balance of an account at some remote terminal (`get_Balance`) and the supply of that value by some account management entity (`provide_Status`):

```
get_Balance(AcctNo:nat, CurrentBal:money [in]);
provide_Status(AcctNo:nat [in], CurrBal:money [out]);
```

When we specify the desired interaction, the implicit flow of information is as follows:

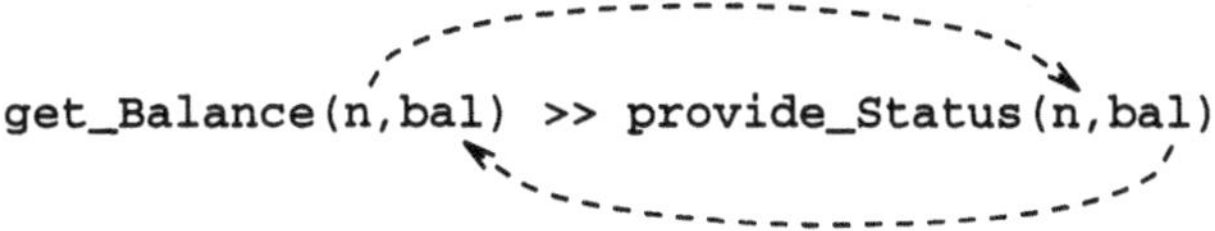

Figure 7.1: Information Flow During an Interaction

The analysis of information flows and possible executions of such interaction specifications is further elaborated in [Saa93, Section 4.2.3] and [HS93].

One way to specify interactions is in TROLL templates of objects that are strongly related to others, i.e. in objects are related by in *is-a*, *role-of*, or *part-of* relationships to other objects. Simple templates are extended by an additional section with the header **interactions**. We do not want to illustrate that at this point but refer the reader to following sections that discuss specifications of strongly related templates.

Formal Syntax

<interactions>	::=	**interactions** *<interaction_def_seq>*`;`
<interaction_def>	::=	*<application_cond><simple_interaction>* | *<simple_interaction><application_cond>* | *<simple_interaction>*
<simple_interaction>	::=	*<event_term> <interact_op> <event_term_list>*
<interact_op>	::=	`>>` | `==`
<event_term>	::=	*<simple_event_term>* | *<selector><simple_event_term>*

$$\text{<}simple_event_term\text{>} \quad ::= \quad \text{<}event_id\text{>}$$
$$| \; \text{<}event_id\text{>}\text{`('}\text{<}data_type_list\text{>}\text{`)'}$$

Semantics

The semantics of calling specifications is easily explained in terms of implications between occurrence predicates for events. Let e_1 and e_2 be events. A calling specification

$$e_1 \gg e_2$$

is translated into a formula

$$\odot\, e_1 \;\Rightarrow\; \odot\, e_2$$

A conditional interaction clause

$$(e_1 \;\gg\; e_2) \;\textbf{only if}\{\; p \;\}$$

translates into

$$(\odot\, e_1 \Rightarrow \odot\, e_2) \;\Rightarrow\; (P(x_1, ..., x_n) \land \varphi)$$

This formula states that the implication over the event occurrences only holds if the predicate P holds, and φ is the translation of the past tense condition p telling us when P holds (see Section 5.1.2.3) where $x_1, ..., x_n$ are the free variables in φ.

Multicasting is translated into a conjunction of implications. Event sharing results in an equivalence, i.e.

$$e_1 == e_2$$

is translated into a formula

$$\odot\, e_1 \;\Leftrightarrow\; \odot\, e_2$$

A note is in order to justify strict calling. We stated above that the called event may only occur if the called event may occur. Assume we have a simple calling clause

$$\odot\, e_1 \;\Rightarrow\; \odot\, e_2$$

Using the inference rule $Loc\odot\triangleright$ (see page 73) we can deduce that $\triangleright e_2$ holds in the current state, i.e. e_2 must be enabled. If we also deduce that e_2 is not enabled in the same state, the proposition ABS in Proposition 4.3.16 can be applied—we may deduce that $\odot\, e_1 \;\Rightarrow\; \odot\, e_2$ does not hold in the current state. Thus, the calling does not occur.

Formulas describing interactions on the semantical level always appear in the smallest local specification containing the interacting instances. For specializations, this is the specialization aspect, for aggregations it is the aggregation itself.

After having discussed the basic principles being required to specify more complex templates in TROLL, let us now present the different constructions.

7.3 Specialization

The Specialization concept is a well-known concept in knowledge representation and semantic data modeling. From the point of view that is adopted by the approach described in this thesis, a specialization can be characterized as a description of a more detailed aspect of a single conceptual object. The traditional point of view characterizes a specialization as an *is-a* relationship.

A specialization is a *static* concept—a specialized aspect of a conceptual object exists for the whole lifetime of the base aspect if it exists at all. The existence of a specialized aspect only depends on static properties of the base object. Thus, specialization can only be described using *static properties* of the base template.

Let us illustrate the specialization concept with a few examples. One of the most striking examples is the issue of special person aspects, namely woman and man. A man or a woman *is_a* person always, and (usually) this special aspect of a particular person always exists during the existence of the particular person and, moreover, there is only *one* possible special aspect instance for a particular person instance: a person cannot be both a man and a woman. In TROLL, we would specify the following:

```
class Person
  attributes
    Sex:{male,female} [constant]
  ...
end class Person;


class Woman
  specializing Person where Sex = female;
    attributes
      ...
end class Woman;

class Man
  specializing Person where Sex = male;
    attributes
      ...
end class Man;
```

Note that the classification of Person instances is done *data driven*, i.e. it depends on the value of a constant attribute. This is the only way we can specify specialization on the instance level.

Recall that is not allowed to directly change inherited attributes in the specialization (encapsulation). Thus, all effects on the base object that are induced by
events local to the specilized aspect only must be defined by interactions. This is
illustrated by the following example.

Example 7.3.1 In this example we want to specify how events local to the specialization may have an impact on the local state of the base. Assume that in the
specification of the `Account` template (Example 6.1.8) we specified an attribute

```
Type:checking,saving [constant];
```

Consider the specification of a `SavingsAccount` being a specialization of the class
`Account`:

```
class SavingsAccount
  specializing Account where Type = saving;
    attributes
      Interest_Rate:real;
    events
      ...
      pay_interest;
    interaction
      pay_interest >> credit(Balance * Interest_Rate / 100);
    ...
end class SavingsAccount;
```

Of course this specification does not exactly reflect the real world since interest is
not paid for the balance at a given point in time, but this example is considered
sufficient for what we want to say.

The crucial point is the interaction specified in the **interaction** section: when the
event `pay_interest` occurs in the specialized aspect, this implies the synchronous
occurrence of an event of type `credit` where the actual parameter is a term denoting the actual interest to be paid for that savings account.   ■

Usually, a specialized aspect does impose further constraints. Since the signature
and the specification of the base aspect are included in the specialized aspect, we
may further restrict properties specified for the base.

Example 7.3.2 A `SavingsAccount` may not be in the red at any time. Thus, in
the specification of `SavingsAccount` we specify the following constraint:

```
class SavingsAccount
  specializing Account where Type = savings;
    ...
    constraints
      Balance >= 0;
    ...
end class SavingsAccount;
```

■

Formal Syntax

The production rule generating a class specification must be expanded by the (optional) symbol *<specialization>* which is defined as follows:

<*specialization*>   ::=   **specializing** <*class_id*>
                                  [**where** <*state_formula*>]

A contextsensitive condition is that the **where** condition may only be a state formula over constant terms.

Semantics of Specialization

The OSL specification of a specialization can easily be constructed along the lines of specialization being a specification morphism (Definiton 4.3.23). We must obey, however, that we may not have a specialization aspect for every base instance because of the **where** condition. Thus, we must restrict every formula concerning the specialization aspect by the specialization condition: a formula local to the specialization is valid if the specialization condition is fulfilled for a particular instance.

For the translation of the `SavingsAccount` specification, this results in formulas of the following form:

- For the constraint `Balance >= 0` we have the formula

$$\triangleright \texttt{Type}(\texttt{savings}) \Rightarrow \Box(\triangleright \texttt{Balance}(m) \Rightarrow (m \geq 0))$$

Please note that a specialization introduces a subsort relationship between the identifier sorts.

Let us illustrate the semantics of specializations by the example of the specification of `SavingsAccount`. The signature is defined as

$$\Theta = (\mathcal{IS}, \alpha, ATT, EVT)$$

where

- $IS = \{|\text{Customer}|, |\text{Account}|, |\text{SavingsAccount}|\}$;

- $\leq_{IS} = \{=_{IS} \cup (|\text{SavingsAccount}|, |\text{Account}|)\}$, i.e. we have a subsort relationship $|\text{SavingsAccount}| \leq |\text{Account}|$;

- $\alpha = |\text{SavingsAccount}|$;

- $ATT = \{\ \ \{\text{No}\}_{\rightarrow \text{nat}}, \{\text{Balance}, \text{Limit}\}_{\rightarrow \text{money}}, \{\text{Holder}\}_{\rightarrow |\text{Customer}|},$
  $\{\text{Red}\}_{\rightarrow \text{bool}}, \{\text{Interest_Rate}\}_{\rightarrow \text{real}}\ \}$

- $EVT = \{\ \ \{\text{open}\}_{\text{nat} \times |\text{Customer}|}, \{\text{close}, \text{pay_interest}\},$
  $\{\text{debit}, \text{credit}, \text{set_limit}\}_{\text{money}}\ \}$

The signature of the local specification of **Account** (see Page 133) is related to the local specification of **SavingsAccount** by an inclusion morphism.

The axioms defined in the specification of **Account** (see Page 133) carry over to the set of axioms in the local specification of **SavingsAccount**. Thus, the set of axioms includes the following formulas:

$$Ax_{\text{SavingsAccount}} = Ax_{\text{Account}}$$
$$\cup \{[\triangleright \text{Type}(\text{savings}) \Rightarrow \Box(\triangleright \text{Balance}(m) \Leftrightarrow (m \geq 0))],$$
$$[(\odot \text{pay_interest} \wedge \triangleright \text{Balance}(m) \wedge \triangleright \text{Interest_Rate}(r))$$
$$\Rightarrow \odot \text{credit}(m * r/100)]\ \}$$

These axioms generate a new theory which refines the theory of the base object.

7.4 Roles

A not so common abstraction mechanism for the modeling of problem domains is the *role concept*. A role can be regarded as a temporary specialization. Thus, it is a dynamic concept. Furthermore, several roles can be played at the same time.

We are regarding a special aspect of a base object, but it is not that easy. The problem is known as the counting problem [WJ91]: consider the role **Passenger** of **Person**. When we want to know the number of **Passengers** that have used a particular transportation means, we are not referring to the number of different **Persons** that have used it but to the number of **Passengers**, where one **person** can be a (*different!*) **Passenger** several times! Moreover, a **Person** can be an **Employee** at the same time (s)he is a **Passenger**, (s)he can even be an **Employee** with several companies at the same time.

The specialization construction as presented in the previous section and Definition 4.3.23 for OSL is not suitable for modeling such situations since it denotes a *static construction* of more specific objects.

For the first example, we assume the following specification of objects of type Person to be given:

```
class Person
  attributes
    Name:string [key];
    Birthdate:date [key];
    Sex:   {female,male} [constant];
    Address:address;
  events
    birth(name:string[in], date:date[in], sex:{female,male}[in]) [birth];
    death [death];
    new_address(a:address [in]);
  effects
      [birth(n,d,s)] Name=n, Birthdate=d, Sex=s;
      [new_address(a)] Address=a;
  obligations
    death;
  end class Person;
```

Consider now the role Customer of Person. A Customer usually has a customer-number, which is attached with the becoming of a Customer. We have to, however, be able to access the properties of a Customer which are properties of the Person playing the role of the customer. This is reflected in the permission for the birth event become_customer of the role Customer: the event may only occur provided the underlying Person-object is born before December 31, 1974 (in this example).

```
class Customer role of Person;
    attributes
      CustomerNo:nat [key constant];
      Since:date [constant]
    events
      become_customer(No:nat [in], when:date [in]) [birth];
      cease [death];
    effects
      [become_customer(n,d)] CustomerNo=n, Since=d;
    permissions
      only if{ Birthdate<=31-12-74 } become_customer;
  end class Customer;
```

Such a specification describes a **Person** playing the role of a **Customer**. When a **Person** plays that role, there may be additional properties that describe this role. We may, however, refer to all properties being valid for the **Person** when we refer to the person playing the role of **Customer**.

In general, a role specification has the following characteristics:

- A role specification optionally introduces a set of new attribute symbols. Since the inherited attribute symbols can be identified by qualifying their names with the name of the base object/class, we have no name conflicts. If there are no name clashes, the qualification can obviously be omitted.

- The role object has birth and optional death events. Birth and death events induce the creation and destruction of a role. Thus an object may play the same kind of role several times. We require the birth and death events either to be inherited from the parent of the role or be locally defined. The two mentioned possibilities reflect that the creation may be under local control (the first case) or due to communication with the environment (the second case). As usual we require the birth event to be the first event of a life cycle and a death event the last event of the life cycle of a role object. The death event is optional, that is, an object may play a role for the rest of its life.

- In general, no overriding is possible. That is, everything which is specified for base objects can only be restricted further in a role specification.

- Valuation rules may be specified as usual. They are restricted to be applied only to *locally* defined attributes. Otherwise the update of an inherited attribute violates the locality of attribute updates. On the other hand, valuation rules may contain inherited events. The role object may suffer from events occurring in the base object and change state according to the state changes in the parent object.

- Attribute values from inherited objects may be read, they are visible. All data terms defined in the role object may therefore contain inherited attribute symbols.

- Constraints may be specified for local and inherited attributes. An object playing a role can thus have a restricted behaviour. Some attribute updates may be allowed, others, violating the newly specified constraints, are forbidden. For example, an employee must not be older than 65 years.

- For permissions we can state the same as for constraints. The possible occurrences of events inherited from the parent object may be further constrained during the time between the birth and the death of the role object. Note, that the possibilities for defining new constraints and permissions for inherited attributes and events respectively, induce a necessary translation process during implementation. We will not deal with this topic in this paper.

- Update of attribute values of the parent object is still possible. For this purpose, inherited events may be called (see Section 7.2). Note that the locality of attribute updates may not be violated in this way since it is definitely the only way to change the state of the parent object.

It is also possible to specify *roles over multiple base objects*. As an example consider the following specification:

```
class TeachingAssistant
  role of Student, Employee
    attributes
       . . . .
    events
       Employee.hired(.....)  [birth];
       Employee.fired(....)  [death];
    ...
```

The restriction is that the base types must have a common ancestor. This is illustrated by Figure 7.2.

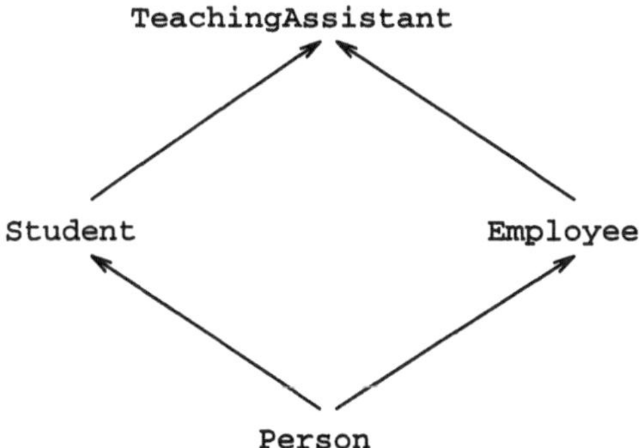

Figure 7.2: Roles over Multiple Base Object Specifications

Formal Syntax

The production rule generating a class specification must be expanded by the (optional) symbol *<role_of>* which is defined as follows:

<role_of>   ::=   **role of** *<class_id>*';'

Semantics of Role Specifications

For the semantics, we have to obey the fact that we have to be able to distinguish different roles of the same base object. Thus, the concept of a specialization is not enough here.

We are regarding a role instance as being an *aggregation* incorporating a base instance and a *role part instance*. This approach fulfills our requirements:

- When base objects play a role, all properties of the base object must be valid for the role object.

- A base object must be able to play the same kind of role several times.

- Each role being played by an instance must be distinguishable from all other roles.

- A base object can play several kind of roles simultaneously.

The fact that we are introducing an object specification of the role part implies that we are able to distinguish roles being played by base objects. That fact that a role playing is represented by an aggregation of the role part and the base part(s) allows to refer to properties of base objects from the role. Since each role has its own role part, several roles may be played by the same base object(s).

A role specification (without the **role of**-section) is translated like a simple object specification into a local signature and a set of local axioms except for those parts of the specification that involve the base object(s). If some of the events of the role are identified with events of the base, we have to generate local copies in the role part signature.

Then we construct the aggregation of the base object(s) with the role part. If there were events identified in the role and the base, we have to specify *interactions* between the corresponding local events.

Finally, if we specified further restrictions on the base in the role specification, we have to construct a specialization over the aggregation were the additional constraints are introduced as axioms. This finally is the representation of a role specification.

Let us first go through a simple example. Consider the specifications of **Person** and **Customer** from above. The **Customer** specification *in isolation* is translated into the OSL instance specification $\tau_{\texttt{CustomerPart}} = (\Theta_{\texttt{CustomerPart}}, Ax_{\texttt{CustomerPart}})$. The signature

$$\Theta_{\texttt{CustomerPart}} = (\Sigma_{,\texttt{CustomerPart}}, \alpha_{\texttt{CustomerPart}}, ATT_{\texttt{CustomerPart}}, EVT_{\texttt{CustomerPart}})$$

has the following components:

$$\Sigma_{,\texttt{CustomerPart}} = \{|\text{CustomerPart}|\}$$

$$\alpha_{\texttt{CustomerPart}} = |\text{CustomerPart}|$$

$$ATT_{\texttt{CustomerPart}} = \{\text{CustomerNo}_{\text{nat}}, \text{Since}_{\text{date}}\}$$

$$EVT_{\texttt{CustomerPart}} = \{\text{become_customer}_{\text{nat,date}}, \text{cease}\}$$

The axioms in $Ax_{\texttt{CustomerPart}}$ can be obtained along the line of the translations described in Chapter 6. The only exception is that formulas that involve symbols that are not in the local signature $\Theta_{\texttt{CustomerPart}}$ cannot be in $Ax_{\texttt{CustomerPart}}$. Thus, we do not have in $Ax_{\texttt{CustomerPart}}$ a formula representing the permission

only if$\{$ `Birthdate <= 31-12-74` $\}$ `become_customer`;

since the attribute symbol `Birthdate` is not local to `CustomerPart`.

The set of axioms of `CustomerPart` includes the following:

$$\Box\,[\odot\,\text{become_customer}(n) \Rightarrow \bigcirc(\Box\neg \vartriangleright \text{become_customer}(n')] \quad (1)$$
$$\Box\,[\odot\,\text{become_customer}(n) \Rightarrow \bigcirc(\vartriangleright \text{CustomerNo}(n))] \quad (2)$$
$$\vartriangleright \text{CustomerNo}(n) \Rightarrow (\bigcirc(\vartriangleright \text{CustomerNo}(n)))\,\mathcal{W}\,death \quad (3)$$

The **Person** specification is translated into $\tau_{\texttt{Person}} = (\Theta_{\texttt{Person}}, Ax_{\texttt{Person}})$ where

$$\Theta_{\texttt{Person}}(\Sigma_{\texttt{Person}}, \alpha_{\texttt{Person}}, ATT_{\texttt{Person}}, EVT_{\texttt{Person}})$$

such that

$$\Sigma_{\texttt{Person}} = \{|\text{Person}|\}$$

$$\alpha_{\texttt{Person}} = |\text{Person}|$$

$$ATT_{\texttt{Person}} = \{\text{Name}_{\text{string}}, \text{Birthdate}_{\text{Person}}, \text{Sex}_{\{\text{female,male}\}}, \text{Address}_{\text{address}}\}$$

$$EVT_{\texttt{Person}} = \{\text{birth}_{\text{string,date,}\{\text{female,male}\}}, \text{death}, \text{new_address}_{\text{address}}\}$$

The set of axioms for the specification of the class **Person** includes the following formulas:

$$\Box\left[\odot\,\mathtt{birth}(n,d,s)\Rightarrow\bigcirc(\Box\neg\,\triangleright\,\mathtt{birth}(n',d',s'))\right] \quad (1)$$
$$\Box\left[\odot\,\mathtt{birth}(n,d,s)\Rightarrow\bigcirc(\triangleright\,\mathtt{Sex}(s))\right] \quad\qquad (2)$$
$$\triangleright\,\mathtt{Sex}(s)\Rightarrow(\bigcirc(\triangleright\,\mathtt{Sex}(s)))\,\mathcal{W}\,death \quad\qquad (3)$$

$$\cdots$$

The template signature for the specification of the aggregation of the parts **Person** and **CustomerPart** looks as follows:

$$\Theta_{(\mathtt{Person},\mathtt{CustomerPart})}=(\mathcal{IS},\alpha,ATT,EVT)$$

such that

$$IS=\{|\mathtt{Person}|,|\mathtt{CustomerPart}|,|(\mathtt{Person},\mathtt{CustomerPart})|\}$$

$$\alpha=|(\mathtt{Person},\mathtt{CustomerPart})|$$

$$ATT=\;\{\mathtt{Name_{string}},\mathtt{Birthdate_{date}},\mathtt{Sex_{\{female,male\}}},\mathtt{Address_{address}},$$
$$\mathtt{CustomerNo_{nat}},\mathtt{Since_{date}}\}$$

$$EVT=\;\{\mathtt{birth_{string,date,\{female,male\}}},\mathtt{death},\mathtt{new_address_{address}},$$
$$\mathtt{become_customer_{nat,date}},\mathtt{cease}\}$$

Thus, since there are no name clashes that have to be resolved to achieve the injection being a requirement for an aggregation. Thus, the mappings $\sigma_{OBS}(\mathtt{Person})$, $\sigma_{EVT}(\mathtt{Person})$, $\sigma_{OBS}(\mathtt{CustomerPart})$, and $\sigma_{EVT}(\mathtt{CustomerPart})$ are just identities.

The set of axioms $Ax_{(\mathtt{Person},\mathtt{CustomerPart})}$ includes all local axioms in $Ax_{\mathtt{Person}}$ and $Ax_{\mathtt{CustomerPart}}$.

Now recall that we referred to an attribute of the base template **Person** in the template specification **Customer**. This results in an additional axiom that is not part of the aggregation. Thus, we have to *specialize* the aggregation (**Person,CustomerPart**) in order to be able to state the corresponding axiom.

On the signature level, we do not have to extend the signature of (**Person,CustomerPart**). The set of axioms $Ax_{\mathtt{Customer}}$ is obtained by adding the formula

$$\Box\,(c.\triangleright\,\mathtt{become_customer}(n)\Rightarrow c.\triangleright\,\mathtt{Birthdate}(31-12-74))$$

to the set of axioms $Ax_{(\mathtt{Person},\mathtt{CustomerPart})}$.

In the presence of multiple inheritance, we proceed along the same lines. Due to the requirement that we have to have a common ancestor in the inheritance graph we have a *common part*. This common base specification must not be replicated. Thus, we are constructing the aggregation of all role parts by considering the non-structured parts only.

For the `TeachingAssistant` example, this is illustrated by Figure 7.3.

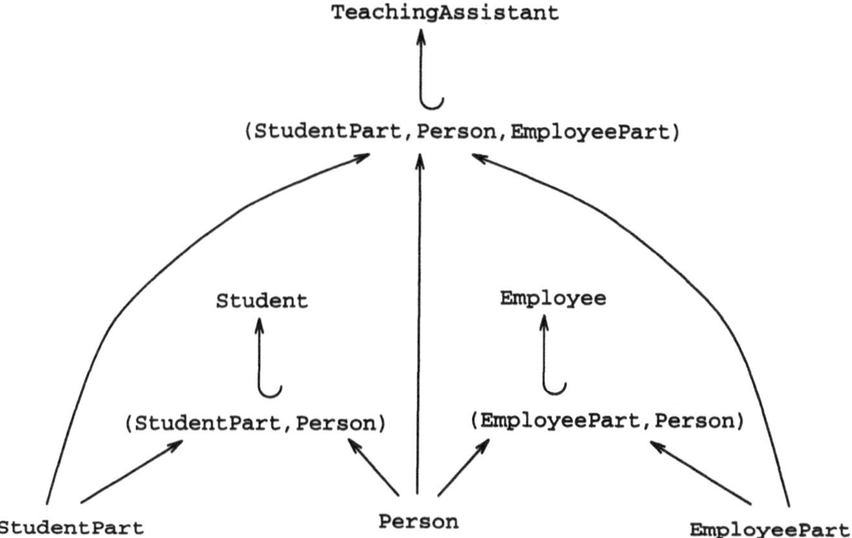

Figure 7.3: Semantic Structure of Roles over Multiple Base Objects

With these considerations we want to proceed to the specification of composite objects.

7.5 Composite Objects

This section deals with the specification of objects that are composed from parts, i.e. from subobjects. In our approach, such objects are called *composite objects*. Composite objects occur frequently in problem domains or dynamic object systems. It is important to provide language features for the specification of composite objects. The concept of constructing objects from other objects has been discussed already under the notion of *aggregated objects* in the presentation of OSL.

In [JSHS91, HJS92], we have presented several language constructs to specify different types of composite objects: static and dynamic composite objects as well as composite objects with local components. We are of the opinion that some of the language constructs can be eliminated or merged.

When we look at domains for information systems, we basically can identify two types of composite objects:

- Objects with *dependent local* components—the structure and behavior of components is globally known, but the existence of component instances depends on the existence of the composite object as a whole, i.e. components cannot exist if they are not a component.

- Objects that *share parts*, i.e. that interact through parts. This requires that the attributes and events of the shared component are accessible within the composite object. Moreover, the component instances may exist independently from the composite object.

- Objects that may *alter their composition*.

In the data modeling community, the first two types of composite objects have been discussed as *disjoint* and *non-disjoint* composite objects [BB84]. The approaches, however, mainly discussed *static* composite objects, i.e. composite objects of which the composition does not change over time. An approach to integrate the concepts of disjoint and non-disjoint composite objects with changing composition was presented in [HJS92].

Please note that in contrast to semantic data modeling approaches that support aggregation here we have to take special care of the *behavior* which is *composed* from the behaviors of the components.

In this version of the language, we restrict ourselves to the following concepts:

- Structure and behavior of component objects are specified globally like any other object or class. Thus, our language version does not include the possibility to specify templates *locally* to other templates.

- The composition may be described by *predicates over constant values* or may be *manipulated by special events*.

- Component names are *local names*. They can be used like class names as described in Section 7.1.

- As structuring concepts for composing we allow for *single components*, for *sets of components*, and for *lists* of components.

- Components may be declared to be *dependent*. A dependent component can only exist if the composite object of which it is a component exists.

This restriction has been introduced among other reasons to reduce the number of language constructs in TROLL. During the development of example specifications we realized that the language should only provide a single language construct to describe composite objects. This language construct, however, then must have a number of options to cover the alternatives above.

In general, we can state the following for composite objects in our model:

- Attribute values of the embedded object may only be altered by events local to it although they may be read inside the enclosing object.

- Projecting a life cycle of a composite object to the events local to a component always yields a life cycle of the component. Note that there may be life cycles of a component that cannot be obtained by projection, i.e. those ones that are not allowed *in the context* of the composite object.

- Interactions do not cause any violation of the life cycle specification of both objects (see below).

Let us now introduce the language features to specify composite objects using some examples.

Example 7.5.1   In this example, we specify a bank object. A bank includes all the accounts maintained by the bank, and they have a manager. The accounts of a bank are modeled as a set of components of type `Account`, whereas the manager is modeled as a single component of type `Employee`. We are only specifying a part of the local signature here, namely the part that introduces components.

```
object Bank
  components
    -- all existing accounts are local, i.e. they may
       not be components of other objects
    Accounts:SET(Account) [local];
    -- managers may change - dynamically...
    Manager:Employee;
  ....
end object Bank;
```

The phenomenon of **Bank** we are specifying in the example above has the following components:

- A *set of* `Account` *instances* and

- a component `Manager` of the class `Employee`. It is allowed to let the component be undefined, i.e. we may have banks which currently do not have a manager, but if they do, it is *at most one* component object.

The set of accounts being components of a composite object in the class `Bank` in the example above is declared to be *local*. Thus, there may be no instance in the class `Account` that is shared between instances of any class. A particular instance in the class `Account` is incorporated in at most one composite instance of class `Bank`. Please note that the structure and the characteristic behavior of an `Account` object is known globally, whereas the actual state and actual behavior of a particular instance is only known inside the composite object that incorporates that component.

The component `Manager` is declared to be a single instance of type `Employee`. When we declare a component to be a single instance, then we allow for situations where there is actually no component incorporated in the current state. Furthermore, we may change the component dynamically during the lifetime of the composite instance. These changes of the composition of a composite instance must be described by *events*.

In TROLL, the specification of a composite object by default denotes the specification of a dynamic composite object. This is a change compared to the previous version TROLL91 [JSHS91, HJS92], where we made an explicit distinction between static composite objects and dynamic composite objects.

Single Components

The following implicit attributes are declared for single components:

- <*comp_name*>_ASSIGNED:bool
  has the value **true** if a component is assigned and **false** if no component is assigned currently.

- <*comp_name*>_ID: | <*comp_class_id*> |
  is an *optional* attribute of which the identifier of the component is the value if the component is assigned.

The following implicit events for manipulating the composition are generated with the declaration of a single component:

- <*comp_name*>_ASSIGN(New_Comp: | <*comp_class_id*> | **[in]**)
  must be used to assign a component to a composite object. The instance

of class <*comp_class_id*> that is identified by the identifier being the actual parameter of the event becomes the new component of the composite object. If a component was defined for the composite object, that component is replaced by the new one. The effects of the occurrence of such an event are that the attribute <*comp_name*>_ASSIGNED has the value **true** and the attribute <*comp_name*>_ID is set to the identifier of the new component.

- <*comp_name*>_REMOVE(Which_Comp: | <*comp_class_id*> | **[in]**)
  is used to remove a component from the composite object.

Example 7.5.2 For the Bank object (Example 7.5.1), we can assume the following implicit specifications that are induced by the component declaration of the single component **Manager**:

- We have the following implicitly declared attributes:

  ```
  Manager_ASSIGNED:bool;
  Manager_ID:|Employee|;
  ```

- The implicitly declared events are

  ```
  Manager_ASSIGN(New_Comp:|Employee| [in]),
  Manager_REMOVE(Which_Comp:|Employee| [in]);
  ```

- For the implicitly declared events we have the following specification sentences defining their effects:

  ```
  [Manager_ASSIGN(e)]Manager_ASSIGNED= true, Manager_ID=e;
  [Manager_REMOVE]Manager_ASSIGNED= false;
  ```

- The implicit events cannot always occur. The following permissions are implicitly specified:

  ```
  only if {not Manager_ASSIGNED or e != Manager_ID} Manager_ASSIGN(e);
  only if {Manager_ASSIGNED} Manager_REMOVE;
  ```

Component Sets

For sets of components, the following signature elements are implicitly defined:

- $<comp_name>$_COMP_IDs:**set**($|<comp_class_id>|$)
  is an attribute over possible sets if identifiers of components. If an instance is a component of the composite object, then its identifier is in the set which is the current value of this attribute.

To alter the set of components the following events are implicitly declared for the composite object:

- $<comp_name>$INCLUDE(New_Comps: **set**($|<comp_class_id>|$) **[in]**)
  denotes the inclusion of the instances identified by the actual parameters in the set of type **set**($|<comp_class_id>|$) into the composite object.

- $<comp_name>$REMOVE(Which: **set**($|<comp_id>|$) **[in]**)
  denotes that the instances identified by the actual parameter of type **set**($|<comp_id>|$) are removed from the set of components. Please note that the identifier type $|<comp_id>|$ is a subtype of the identifier type $|<comp_class_id>|$ denoting the set of actual components.

Example 7.5.3 Consider the **Bank** again from Example 7.5.1 and let us illustrate the implicit specification elements.

The implicit attribute is

- Accounts_COMP_IDs:**set**($|$Account$|$);

Its value is manipulated according to the following implicit effect specifications:

```
[Accounts_INCLUDE(A)]Accounts_COMP_IDs=insert(A,Accounts_COMP_IDs);
[Accounts_REMOVE(A)]Accounts_COMP_IDs=remove(A,Accounts_COMP_IDs);
```

As permissions for the implicitly defined events we have

```
only if{(a in Accounts_COMP_IDs) eq not(a in A)} Accounts_INCLUDE(A); -- (1)
only if{(a in Accounts_COMP_IDs) eq (a in A)} Accounts_REMOVE(A); -- (2)
```

The rule says that (1) only accounts that are currently not components of the complex object **Bank** can become components and (2) that only accounts that are currently components of **Bank** can be removed.

Component Lists

For lists of components, the following attributes are implicitly defined:

- $<comp_name>$_COMP_IDs : list (| $<comp_class_id>$ |)
  is an attribute over possible lists of identifiers of components. If an instance
  is a component of the composite object on a position i, then its identifier is
  in the list which is the current value of this attribute at position i.

The composition of a composite object having a component list can be altered
by employing one of the following events:

- $<comp_name>$APPEND(New_Comp: | $<comp_class_id>$ | **[in]**)
  denotes that the instance identified by an identifier of the appropriate type
  is appended to the list of components.

- $<comp_name>$REPLACE(Pos:nat **[in]**,New_Comp: | $<comp_class_id>$ | **[in]**)
  replaces the component instance at position Pos in the list of components by
  the instance identified by the the identifier New_Comp.

- $<comp_name>$REMOVE(Which: | $<comp_id>$ | **[in]**)
  is used to remove the component identified by the actual parameter Which
  from the list of components. The position that was occupied by the removed
  component remains empty.

- $<comp_name>$ERASE(Pos:nat **[in]**)
  denotes the event of removing the component at position Pos from the list of
  components. The position remains empty.

Obviously, the treatment of single and set components is similar to the treatment
of complex data types of that structure. We will not illustrate lists of components
further here.

Let us now present a specification of a composite object to explain the referencing
of components inside the composite object and to illustrate the specification of
interactions between components.

Example 7.5.4 The bank we specify in this example is just a rudimentary bank
which just maintains accounts. Accounts are opened and closed by the bank. The
opening of an account requires an external communication, which is omitted here.
If an account has been opened, there is a commitment to include the just opened
account into the set of accounts maintained by the **Bank**. If an account is closed,
the component is destroyed and removed from the set of accounts maintained by
the **Bank**. This is specified as an interaction.

```
object Bank
  components
    -- all existing accounts are local, i.e. they may
       not be components of other objects
    Accts:SET(Account) [local];managers may change - dynamic...

  events
    establish [birth]; close_down [death];
    open_account(No:nat [in]);
    close_account(No:nat [in]);
  permissions
    only if{has occurred(open_account(n))} close_account(n);
  commitments
    only if{occurs(open_account(n))} Accts_INSERT(Account(n));
  interaction
    close_account(n) >> Accts(n).close, Accts_REMOVE(Accts(n));
  end object Bank;
```

Please note how referencing takes place inside the complex object. We can use the
term

$$Accts(n)$$

to refer to *an actual component* with a key attribute value n.   ∎

Formal Syntax

<component_section>	::=	**components**
		<comp_spec_seq>';'
<comp_spec>	::=	*<single_comp_spec>*
		| *<comp_set_spec>*
		| *<comp_list_spec>*
<single_comp_spec>	::=	*<comp_id>*':'*<class_id>*
<comp_set_spec>	::=	*<comp_id>*':'**SET**'('*<class_id>*')'
<comp_list_spec>	::=	*<comp_id>*':'**LIST**'('*<class_id>*')'

Please note that we did not include the predefined symbols in the formal syntax.

Semantics of Composite Object Specifications

Intuitively, the semantics of a composite object is based on the aggregation of all possible components. This, however, would not be an adequate representation for dynamic composite objects that may alter their composition in TROLL. Therefore, we must insert another level: actual components are regarded to be *roles* of the component objects; whenever an instance is currently a component of a composite object, it plays the role of a component. Thus, we have to define the additional properties of an instance being a component.

Let us first elaborate on the semantics of composite objects with single and set-valued components. We have a four-level composition which is illustrated by Figure 7.4.

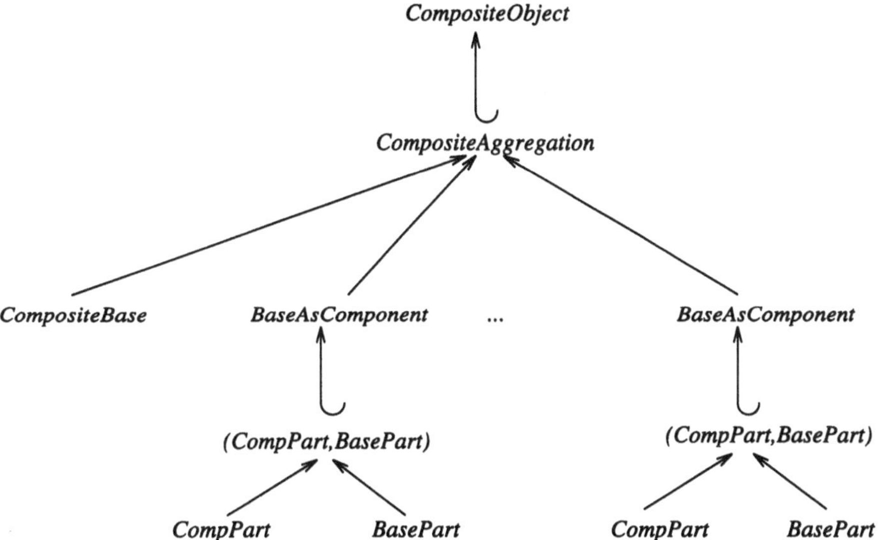

Figure 7.4: Construction of a Composite Object Specification

On the lowest level we have the base components (*BasePart*) and the part that describes that additional properties of base components in the role of a component (*CompPart*). Over these parts we construct in the usual manner (see Section 7.4) a template characterizing the base playing the role of a component (*BaseAsComp*). For each *possible* component a role template is incorporated in an aggregation representing the base of the composite object (*CompositeAggregation*). This construction is similar to the one constructing the base of a class (see Section 6.3).

Thus, a component (be it a single or a set component) is a collection similar to a class. In the *CompositeAggregation* template we furthermore incorporate the local specification of properties of the composite object as a whole (*CompositeBase*). This template is obtained as usual by translating only the local properties of a composite object specification in TROLL. Since we have additional constraints that are only valid for the composite object as a whole, we have to specialize the *CompositeAggregation* into the template specification representing the entire composite object (*CompositeObject*).

Let us now turn to the general structure of the *CompPart*, i.e. to the specification of the additional properties of an object when playing the role of a component. We are only regarding single components and sets of components, which implies that the additional properties are just the events for starting or ceasing to play the component role. The signature of the template specification *CompPart* is defined as follows:

$$\Theta_{CompPart} = (\mathcal{IS}_{CompPart}, \alpha_{CompPart}, ATT_{CompPart}, EVT_{CompPart})$$

where

- $IS_{CompPart} = \{|\texttt{CompPart}|\}$;

- $\alpha_{CompPart} = |\texttt{CompPart}|$;

- $ATT_{CompPart} = \emptyset$ and

- $EVT_{CompPart} = \{\texttt{bc_component}, \texttt{not_component_anymore}\}$

Over the local template specification for the component objects (*BasePart*) and the local template *CompPart* we construct in the usual way the template *BaseAsComp* as described in Section 7.4 for role objects.

A composite object always has a local template which at least contains signature elements and axioms describing the properties of the composite object as a whole. Besides the implicitly generated signature elements and axioms we may have local signature elements (like the event symbols in Example 7.5.1) and even local constraints or local behavior specifications. These local properties are translated in the usual way into a template specification *CompositeBase*. Let us describe here the implicitly generated signature elements for single components and sets of components

The signature of the local part of a composite object with a single component has the following general form:

$$\Theta = (\mathcal{IS}_{CompSingleBase}, \alpha_{CompSingleBase}, ATT_{CompSingleBase}, EVT_{CompSingleBase})$$

where

- $ATT_{CompositeSingleBase}$ includes at least the following elements:

  - $ASSIGNED_{bool}$

  - $ID_{|comp_id|}$

- $EVT_{CompositeSingleBase}$ includes at least the following elements:

  - $ASSIGN_{|comp_id|}$

  - $REMOVE_{|comp_id|}$

The signature of the local part of a composite object with a set of components has the following general structure:

$$\Theta = (\mathcal{IS}_{CompositeSetBase}, \alpha_{CompositeSetBase}, ATT_{CompositeSetBase}, EVT_{CompositeSetBase})$$

where

- $ATT_{CompositeSetBase}$ includes at least the following elements:

  - $COMP_IDs_{\mathbf{set}(|...|)}$

  - $CARD_{nat}$

  - $IN_{|...|,bool}$

- $EVT_{CompositeSetBase}$ includes at least the following elements:

  - $INCLUDE_{|comp_id|}$

  - $REMOVE_{|comp_id|}$

The formulas in a composite object are obtained from translating local formulas and mapping formulas local to components into the local specification describing the composite object. This can be done in the usual manner and shall not be illustrated in detail here.

A word is in order about referencing components inside a composite object. There is an implicit attribute $KeyMap(s_1,...,s_k,|Class|)$ as explained in Section 7.1 for each component. The selector function is defined as

$$<component>(\vec{x}) = \begin{cases} \perp & \text{if } \triangleright COMP_IDs(A) \wedge Class(\vec{x}) \notin A \\ o & \text{if } \triangleright KeyMap(\vec{x}, o) \end{cases}$$

This concludes the presentation of language features to describe strongly related aspects. In the following chapter we describe how separate local specifications can be put into a system context.

8 Specification of Systems

In this chapter, the language constructs for the specification of systems that are *constructed* from larger parts are presented. At first sight, we may claim that we can just use the constructs for the description of complex objects introduced in the previous chapter since a system can be regarded as a complex object with just the existing objects as components.

This is a nice perspective from a' semantical point of view since we do not need to introduce additional features. From the development point of view, however, this seems not adequate for the following reasons:

- Systems to be specified nowadays tend to be very complex, so that there is hardly a chance to understand the different relationships between the components as a whole. Thus, it is at least very difficult to specify the system as one complex object since then one has to specify all the relationships between otherwise unrelated objects at a central place in this complex object specification.

- The task of specifying complex systems today is a task for a number of teams, not for a single one. Thus, different teams work on different components of the system simultaneously. It is much easier to integrate the work of the different teams if there are *interfaces* that do not force the integrator (team) to fully understand the subsystems in order to be able to connect them to form a complete system. The same reason also makes a large-grained structuring mechanism for specifications above the initial structuring in objects desirable.

- In the phase of conceptual modeling, the system is specified as a closed system (see Page 2). Thus, there may be components which are not computerized or which are not even computerizable at all, but which have relationships with other components that are computerized. Then clean interfaces between them are very helpful since they provide that basis for the development of interfaces to users or external components (like e.g. sensors).

These arguments support our view to provide language features that support the development of large, complex, structured systems. In this chapter, we will introduce such features: *Relationships*, *Interface Specifications*, and additional features for the specification of *Object Communities*.

8.1 Relationships

Relationships in semantic data models and object models usually describe some form of (undirected) association between instances of (several) classes. Relationships originate from the Entity-Relationship Model which explicitly supports the distinction between the *local description* of abstract objects and the relationships *between* them.

In object-oriented models, this view has often been neglected since it did not make sense in an implementation-oriented point of view. Traditional data models represented relationships either by values (in the relational model) or by pointers or references (in the network and hierarchical models).

So far, relationships have been a concept of the *structural description.*

Relationships between dynamic objects are, however, not as simple as relationships in e.g. the Entity-Relationship model—here, we must be able to specify interactions between objects and dependencies between objects [JHS92]. Other authors have pointed out the need to have relationships in high-level object-oriented modeling [RBP+91], but those relationships describe only static relationships between object references. We can identify the following reasons for supporting explicit relationships in conceptual object-oriented modeling:

- Global properties should be separated from local properties of objects. This becomes important when we are in the progress of embedding components into a system context.

- We are able to model communication between components and dependencies between components *explicitly*. It is unnatural to bury information about relationships into the related objects.

- Explicit relationships help to keep context-dependent interrelationships out of object descriptions. This supports *reusability* of objects (which now do only contain local descriptions) and make the local assessment or verification of object descriptions (or component descriptions) possible. This clearly helps in reducing the costs to verify or validate system specifications.

- A conceptual model also describes the environment of the system to be implemented. Thus, relationships between objects in the system and objects of the environment can turn out to be descriptions of the functionality of interfaces to the system. Interfaces are introduced by the implementation of the system and are thus not a part of the initial conceptual model of the organization.

In TROLL, the relationship construct supports two kinds of non-localized relationships between objects:

- *Global Constraints* state integrity rules between objects. Such constraints cannot be specified local to the participating objects.

- *Global Interactions* describe communications between separated instances. According to the locality principle in TROLL, this cannot be specified directly between separately defined templates since the signature elements are not globally known.

Let us illustrate the use of the relationship construct with a few examples.

Example 8.1.1 The following example describes communication relationships between an ATM and the corresponding Bank.

In the specification of an extended bank **Bank2** we assume the following events to be declared:

```
object Bank2
   . . .
   events
      . . .
      check_card(No:nat [in], PIN:nat [in]);
      card_OK; card_NOK;
      process_TA(AcctNo:nat [in], Amount:money [in]);
      TA_OK; TA_failed;
      . . .
end object Bank2;
```

The events denote the receiving of a request to check a card and personal identification number (PIN) (`check_card`), the possible results of this check (`card_OK`, `card_NOK`), the receiving of a request to process a remote transaction (`process_TA`), and indications whether a transaction was successful or failed (`TA_OK`, `TA_failed`).

The relationship `RemoteTransaction` specifies a communication relationship between instances of the class `ATM` and the (separately specified) object `Bank2`. Interaction is specified by relating events of both specifications.

```
relationship RemoteTransaction between Bank,ATM
  interaction
    -- Card checking business
    ATM(n).check_w_bank(a,p) >> Bank2.check_card(a,p);
    only if{previous(occurs(ATM(n).check_w_bank(a,p)))}
        Bank2.card_OK >> ATM(n).card_OK;
    only if{previous(occurs(ATM(n).check_w_bank(a,p)))}
        Bank2.card_NOK >> ATM(n).card_not_OK;
    -- Bank transaction business
    ATM(n).issue_TA(n,m) >> Bank2.process_TA(n,m);
    only if{previous(occurs(ATM(n).process_TA(a,m)))}
        Bank2.TA_OK >> ATM(n).TA_ok;
    only if{previous(occurs(ATM(n).process_TA(a,m)))}
        Bank2.TA_failed >> ATM(n).TA_failed;
  end relationship RemoteTransaction;
```

The first three calling sentences concern the checking of a cash card inserted
into an ATM. The ATM sends a message to the `Bank2` object by calling the event
`check_card` in the Bank2. The actual parameters of the event `check_w_bank(a,p)`
are sent to the `Bank2` by instantiating the parameters of the event `check_card`
with the same values. The `Bank2` sends messages back (in the next state) by
calling events that model the receiving of a positive or negative result of the card
checking (`card_OK` or `card_not_OK`). Please note that the precondition is necessary
to bind the value of the variable n in order to call the event in the right instance
of the class `ATM`. The communication is described analogously for the carrying out
of a bank transaction.

Formal Syntax

$<rel_def>$ **::=**   **relationship** $<rel_id>$ **between** $<item_list>$
                           $[<constraint_section>]$
                           $[<interactions>]$
                           end relationship $<rel_id>$`;`

Semantics

Relationships in TROLL model relationships between otherwise unrelated components, i.e. those components are a priori not related through a specialization, a
role, or in a complex object. This can be modeled rather straightforwardly using
the concepts of global interactions in the Object Specification Logic (see Definition

4.3.26). A bit more difficult is the situation with global constraints since additional constraints over an aggregation can be introduced only through a specialization of the aggregated object.

In general, relationships are modeled in OSL over the aggregation of the participating local specifications. Thus, we are sacrificing the separation on the semantical level and consider the context including the related objects as a complex object.

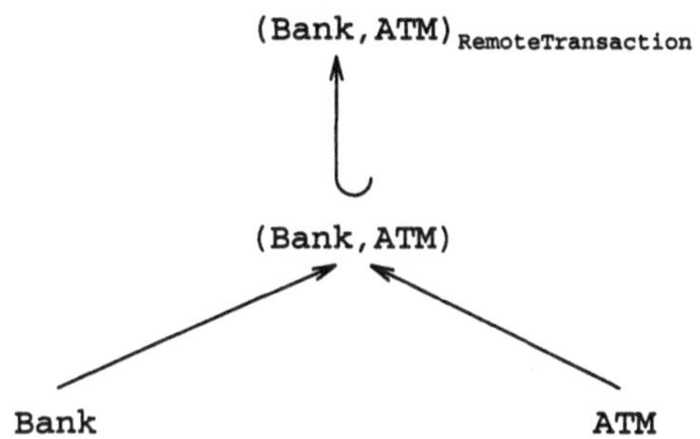

Figure 8.1: Semantics of Relationships

Interactions can be specified already in the aggregation over the participating objects.

8.2 Interfaces

Object systems may be itself embedded in a larger system, they may model subsystems. In order to hide the complexity of the subsystem it should be possible to define *external interfaces* for an object systems in order to explicitly provide views and to restrict access to properties of the encapsulated system. This corresponds to the principle of separation of concerns.

The usefulness of interfaces (there called *abstractions*) has already been pointed out in [ES91]. The process model underlying objects allowed for the spontaneous occurrence of changes in attribute values, thus dropping the frame rule. Abstractions are full-fledged object aspects where certain actions and observations are hidden. Defining abstractions over objects means to hide details, but not forgetting them—their effects may be visible at the interface.

In TROLL91 [SJE91, JSHS91, SJ92] we introduced interfaces as another concept of the language. The object model we used was, however, not capable of handling non-deterministic changes in attribute values. Thus, interfaces were introduced as a mere *syntactical* concept, the models were the instances underlying the interface.

At the level of the interface, access is only possible through the signature elements
defined for the interface—the primary purpose of interfaces is thus access and
visibility restriction.

In the present framework using OSL for the semantics, spontaneous changes in
the possible observations are not possible. Thus, interfaces are not objects but only
restricted views on objects, where only certain properties are visible. Technically,
an interface defines a set of formulas that are also formulas in the theory of the
underlying object. Additionally, we may have formulas that describe additional
properties based on the visible properties at the interface.

8.2.1   Interface Specification

In contrast to the variety of ways to specify interfaces that have been proposed in
[JSHS91] and [Saa93] we only support the specification of *projection interfaces* in
TROLL. A projection interface restricts the set of signature elements of a template
specification, and we may define derived attributes of which the values can be
computed based on the attributes visible in the interface. Let us illustrate this
with examples.

In the first example we specify an interface over an object representing a calendar
showing only the current date to the outside. The `Calendar` is specified as a single
object:

```
object Calendar
  attributes
    Date:date;
  events
    start(When:date [in]) [birth];
    next_day;
  effects
    [start(d)]Date=d;
    [next_day]Date=add_days(1,Date);
  obligations
    next_day;
  commitments
    next_day only if{occurs(next_day) or ((exists d:date)(occurs(start(d))))};
end class Calendar;
```

The interface `Current_Date` over `Calendar` only has one attribute: the current
date. The interface is described by the following specification:

```
interface Current_Date
  encapsulating Calendar;
    attributes
      Date:date;
end interface Current_Date;
```

At the interface we are not able to deduce how the value of the attribute is determined. We may, however, deduce that if the current date changes, then it is increased by one day. This will be explained below.

We may also define interfaces for instances of a class. Consider the specification of the class ATM (Example 6.1.7 on page 129): a user does not have to see the events check_w_bank, card_ok, TA_OK, or issue_TA (denoting the request to check a card issued by the ATM, the reception of a message concerning a valid card and a successful transaction, and the issuing of a request to perform a bank transaction by the ATM) because they do not concern his interaction with the ATM.

The interface would be defined as

```
interface ATM_to_User
  encapsulating ATM;
      ...
    events
       ...
      insert_card(AcctNo:nat [in]);
      input_PIN(PIN:nat [in]);
      card_not_OK;
      input_Amount(Amount:money [in]);
      TA_failed;
      deliver_cash(Amount:money [in]);
end interface ATM_to_User;
```

The interfaces are specified for *instances* of the class ATM, not for the class itself. We may use the same reference mechanism for interfaces on class instances that was defined for collections in Section 7.1. Thus, if n is a natural number, then

$$\text{ATM_to_User}(n)$$

references the interface on top of the instance $\text{ATM}(n)$.

Formal Syntax

$<interface_spec>$ ::=   **interface** $<interface_id>$
            encapsulating $<class_id>$
            $[<attribute_section>]$
            $[<event_section>]$
            $[<derivation_section>]$
            end interface $<interface_id>$`;'

As context-sensitive properties we claim the following:

- If an attribute symbol of the interface is not an attribute symbol of the encapsulated object, this attribute symbol must be a derived attribute.

- No additional event symbols may be declared in the interface.

Semantics

An interface defines a local signature, which is obtained by restricting the local signature of the underlying specification to the symbols defined in the interface. Please note that the components EVT or ATT of a local signature $\Theta_{\text{interface}} = (\mathcal{IS}, \alpha, ATT, EVT)$ may be empty sets. The semantics of an object specification $\mathbf{T} = (\Theta, Ax)$ is a generated set of formulas, the theory $Ax^{\vdash\Theta}$.

The semantics of an interface is defined as the set of formulas that are built exclusively over the signature elements in $\Theta_{\text{interface}}$. Please note that this is not a theory in the sense of Definition 4.3.14, since it is not closed against the inference rules!

Example 8.2.1 Let us illustrate that with an example. In the theory for the object `Calendar` (see Page 174) we have (among others) the following formulas:

$$\exists d:\ \odot\, \mathsf{start}(d) \Rightarrow \bigcirc(\square(\triangleright\, \mathsf{next_day})) \qquad (1)$$
$$\exists d:\ \odot\, \mathsf{start}(d) \Rightarrow \bigcirc(\odot\, \mathsf{next_day}) \qquad (2)$$
$$\exists d:\ \Diamond(\odot\, \mathsf{start}(d)) \qquad (3)$$
$$\exists d:\ \odot\, \mathsf{start}(d) \Rightarrow \bigcirc(\triangleright\, \mathsf{Date}(d)) \qquad (4)$$
$$\exists d:\ \Diamond(\triangleright\, \mathsf{Date}(d)) \qquad (5)$$
$$\exists d:\ \triangleright\, \mathsf{Date}(d) \Rightarrow \bigcirc(\triangleright\, \mathsf{Date}(d +_{\mathbf{date}} 1)) \qquad (6)$$

For the interface `Current_Date` over `Calendar` we have only formulas that contain symbols defined in $\Theta_{\text{interface}}$, e.g. the following ones:

$$\exists d:\ \Diamond(\triangleright\, \mathsf{Date}(d)) \qquad (1)$$
$$\exists d:\ \triangleright\, \mathsf{Date}(d) \Rightarrow \bigcirc(\triangleright\, \mathsf{Date}(d +_{\mathbf{date}} 1)) \qquad (2)$$

∎

8.2.2   Using Interfaces

Interfaces can be imported into template specifications similar to components. In contrast to components interfaces do not make up complete objects—we only have access through a limited signature and may not be able to construct a complete theory for the resulting template. For structuring a specification as well as for constructing specifications for composite systems, however, the import of interfaces is a very helpful language feature.

As an example consider a class Account which imports the interface showing the current date. In TROLL, the import would be specified as follows:

```
class Account
  importing Current_Date;
    ...
```

We may now refer to the attribute Date of the interface Current_Date, e.g. for a log of transactions for a particular account. This log would be modeled as a list-valued attribute Log. The data type LogEntry would be defined as isomorphic to **tuple**(date,money) with a constructor operation LogEntry:date $\times$ money $\rightarrow$ LogEntry. Every update (credit or debit) would cause a value of type LogEntry to be appended to the Log. An extension of the class specification Account (Example 6.3.1 on Page 136) would be defined by the following specification fragment:

```
class Account
  importing Current_Date; ·
  attributes
      ...
    Log:list(LogEntry);
  events
      ...
    debit(Amount:  money [in]);
    credit(Amount:money [in]);
      ...
  effects
      ...
    [debit(m)] Log=append(LogEntry(Current_Date.Date,-m),Log);
    [credit(m)] Log=append(LogEntry(Current_Date.Date,m),Log);
      ...
  end class Account;
```

For interfaces on classes, we may import only interfaces over selected instances. This may be specified in the **importing** section by a boolean condition over constant properties.

Example 8.2.2 Suppose e.g. that we specified an interface `Acct_Balance` over the class `Account`:

```
interface Acct_Balance
  encapsulating Account;
    attributes
      No:nat;
      Balance:money
end interface Acct_Balance;
```

Suppose further that a particular object `Special_Acct_Service` has access to the accounts with numbers 1 to 1000. This could be specified as follows:

```
object Special_Acct_Service
  importing A in Acct_Balance where (A.No >= 1 and A.No <= 1000);
  ...
end object Special_Acct_Service;
```

A is a variable that takes instances of interfaces. ∎

Formal Syntax

The **importing** clause has the following syntax:

$<import_sec>$ ::=   **importing** $<import_seq>$ '`;`'

$<import>$ ::= $<interface_id>$
                                     | $<sel_import>$

$<sel_import>$ ::= $<obj_var_id>$ **in** $<interface_id>$
                                     [**where** $<state_formula>$]

Semantics

The signature of a local specification is extended by the signature elements of the imported interface. The theory of a local specification is extended by the formulas that are valid for the interface. Please note that the frame rule does not hold anymore for the resulting set of formulas. The semantics of a specification with an imported interface thus cannot be defined within the OSL framework.

8.3 Object Systems

A number of object and class specifications along with relationships and interfaces makes up the specification of an *Object System*. An object system is a unit consisting of coherent interacting objects. Object system specifications are only syntactical barriers aruond a set of related specifications. There is no further semantics intended with an object system specification.

In order to support higher levels of modularity, *system interfaces* can be used. System interfaces encapsulate entire object systems like interfaces encapsulate objects. Like modules in programming languages, system interfaces only export interfaces to the outside, thus controlling access to and visibilty of properties defined locally in object and class specifications. First ideas on modularization can be found in [SJE91].

We are aware of the fact that this can only be a first solution supporting the structured specification of large object systems. Further experiments with sufficient examples have to be carried out to evaluate the proposal.

8.3.1 Object System Specification

To specify an object system, one just has to list specifications of classes and objects.

n a system specification, we can import interfaces from system interfaces (which are described in the next section). These imported interfaces can be used in object specifications as described in Section 8.2.2. We will not elaborate further on this feature since it is just a declaration of interfaces that can be used in object and class specifications within a system specification.

The following conditions must be fulfilled syntactically:

- Classes of which roles and specializations are defined must be specified in the object system specification.

- Classes of which instances can be components of composite objects must be specified.

- Participants in relationships must be specified as objects or classes in the object system specification.

- Each interface imported into an object or class specification must either be defined in the object system specification or it must be imported into the object system specification.

With these rules we ensure that all references to specification elements can be resolved within the object system specification.

An object system specification for our banking world would be specified as follows:

```
object system BankingWorld
  importing Current_Date from Timer_Date;

  object Bank
    . . .
  end object Bank;

  class Account
    . . .
  end class Account;

  ...
end object system BankingWorld;
```

Timer_Date is a system interface to an object system dealing with descriptions of phenomenons related to time and calendars. System interfaces are briefly described in the next section.

Formal Syntax

Specification texts for object system specifications are generated using the following rules:

<system_spec>	::=	**object system** *<system_id>*
		[*<import_seq>*]
		<spec_item_seq>
		end object system *<system_id>*
<spec_item>	::=	*<object_spec>*
		| *<class_spec>*
		| *<relationship_spec>*
<import_spec>	::=	**importing**
		<sys_import_seq>
<sys_import>	::=	*<interface_id_list>* **from** *<sys_interface_id>*

Semantics

The semantics of an object system specification is defined along the lines of a system specification in OSL (see Section 4.3.3.1). A system signature Ψ includes a partially ordered set $\mathcal{IS}$ of identifier sorts such that all possible combinations of sort $\mathsf{tuple}(s_1, ..., s_n)$ for identifier sorts $s_1, ..., s_n$ are included. $\mathcal{IS}$ includes all identifier sorts that are defined by local signatures. Ψ also includes a family of local signatures $\{\Theta^{is}\}_{is \in IS}$ indexed by the identifier sorts in $\mathcal{IS}$. That is, we have for each local identifier sort a local signature.

According to the relationships between signatures, we construct the local specifications over the items defined in an object system specification in TROLL.

- Object specifications are represented by local specifications.

- Class specifications are represented by local instance specifications, by the aggregation over all possible instances, and by the class specification specializing the aggregation.

- Composite object specifications are represented by the local component specifications, by the aggregation of all components, and by the local composite specification over the aggregation.

Furthermore, $\{\Theta^{is}\}_{is \in IS}$ includes all possible aggregations over local signatures. Relationships then are transformed into interactions in the least aggregation incorporating the participating local specifications.

8.3.2 System Interfaces

Object Systems can be related to other object systems through *System Interfaces*. A system interface consists of a number of object interface specifications over the objects and classes defined for an object system. It just encapsulates a whole object system by providing only interfaces to selected components of the encapsulated system. A system can therefore access services of other separately defined systems in the same way as objects may have access to the properties of other objects through interfaces.

In a system specification, we can import interfaces from system interfaces. These imported interfaces can be used in object specifications as described in Section 8.2.2. We will not elaborate further on this feature since it is just a declaration of interfaces that can be used in object and class specifications within a system specification.

Suppose we have an object system `Timer` which (among others) contains the specification of the object `Calendar` (Example 8.2.1 on Page 174). A system interface `Timer_Date` over `Timer` consisting only of the interface `Current_Date` (see Page 175) would be specified as follows:

```
system interface Timer_Date
   encapsulating Timer;

   interface Current_Date
      encapsulating Calendar;
         attributes
            Date:date;
      end interface Current_Date;

   end system interface Timer_Date;
```

Formal Syntax

$$\langle system_interface\rangle \quad ::= \quad \textbf{system interface}\langle sys_interface_id\rangle$$
$$\textbf{encapsulating } \langle system_id\rangle$$
$$\langle interface_spec_seq\rangle$$
$$\textbf{end system interface } \langle sys_interface_id\rangle$$

At this point, all language features of the language TROLL have been presented. We now want to relate the approach to others in the field.

Part III

Discussion

9 Related Approaches

This chapter is about related approaches. In the last years, a number of object-oriented approaches to system modeling have been proposed which put strong emphasis on the specification of behavior. We divide the presentation into two sections:

- In the first section, a number of formal textual languages are discussed. Among these languages are OBLOG [SSE87] and OBL-89 [CSS89] as precursors of TROLL, CMSL [Wie90, Wie91c], TLOOM [Ara91, Ara92], Glider [DDRW91] and its successors Albert [DDR92, DDP93] and Clyder [HRW93].

- In the second section, we are discussing three graphical formalisms: the graphical version of OBLOG [SSG$^+$91], *Object/Behavior Diagrams* [KS90, KS91], and the *Object Modeling Technique* [RBP$^+$91].

We are briefly mentioning other approaches that are related but not closely related.

9.1 Formal Languages for Object-Oriented Modeling and Specification

Lately, a number of proposals for object-oriented conceptual modeling languages that emphasize the modeling of system dynamics have been proposed.

OBLOG and OBL–89

In 1987, the language OBLOG was presented [SSE87]. It was the starting point for the development of a number of languages which were precursors of TROLL. The language was based on a algebraic framework for the dynamics of objects with the components

- a universe of data values, among them complex values that identify objects

- a set A of attributes, a set X of events,

- a set Λ of life cycles, i.e. of admissible finite and infinite sequences of events, and

- a valuation function associating with each finite prefix of a life cycle a value for each attribute $a \in A$.

The basic unit of specification in OBLOG are *object types*. An object type defines a set of potential instances, their structure and admissible behavior. The potential instances are specified by explicitly constructing the name space (along with invariant attributes). The structure is defined by attributes (along with constraints defining the admissible values). The behavior is defined by *events* and *trace specifications* where the admissible sequences of events are specified along with the effects of events on the values of attributes. Interaction between objects are defined by identifying events (please note that the event space is global).

The language OBL-89 [CSS89] was developed on the basis of OBLOG. OBL-89 is one of the ancestors of TROLL and thus shows a lot of similarities in notation. We will just list the most significant differences here:

- Identifier spaces (*surrogate spaces*) are defined directly in object descriptions, i.e. we may specify how a surrogate space is constructed over simple data types.

- In OBL-89, relationships of any kind between objects are specified explicitly by *links*. Links describe morphisms between object descriptions. A link is always described in the source specification of a morphism. A link consists of

  - a *surrogate map* defining the relationship between the surrogate spaces, and

  - a *template map* where maps between event and attribute terms are defined which describe identical attributes and events.

  An object morphism also defines which *instances* will be incorporated. The selection is done over surrogates and constant attributes (which are also modeled as part of the surrogate space).

- OBL-89 does not distinguish between classes and class types. A template along with a surrogate space defines an *object type*. Object types furthermore can be homogeneous (i.e. they are described by a single template) or heterogeneous (i.e. they are described by several templates and thus the instances can be structured differently).

Based on OBL-89, the development of the high-level languages **Oblog**$^+$ [JSS91] and later TROLL [JSHS91] was taken out. These languages in turn were the immediate precursors of TROLL as described in this thesis.

TLOOM

TLOOM (*Temporal Logic* based *Objec-Oriented Model*) is a language for the modeling of dynamic aspects of database applications that combines approaches for the temporal-logic based specification of database behavior and object-oriented approaches [Ara91, Ara92]. TLOOM emphasizes the description of the temporal evolution of object behavior and the temporal composition of object behavior.

The fundamental notions of TLOOM are:

- *Objects* directly reflect real-world objects. An object has a unique identifier and a completely hidden internals state. An *elementary* object is defined independently of other objects, whereas a *composite* object is comprised from dependent component objects.

- *Messages* that can be sent and received from objects allow for interactions between objects. The receiver of a message can react by changing its current state, by sending messages to other objects, or by returning a value to the sender (similar to Actors, described on Page 46. Messages make up the interface of an object. The description of admissible orderings of messages over time is the central task supported by TLOOM.

- *Roles* describe particular aspects or behavior that an object exhibits during a period of time, which coincides with the notion of role in TROLL. An object can play several different roles at the same time, but it cannot play several instances of the same role simultaneously. A role can be compared to a class. Objects then can be instances of several classes and can migrate between classes (in traditional object-oriented terms).

- *Contexts* comprises roles and public constraints. Each object can only be in one context.

- *Composite contexts* comprise several objects with their roles. Public constraints here describe the global relationships between the components and their local behaviors.

TLOOM does not support the notions of specialization/generalization but claims that these concepts can be adequately represented by roles.

TLOOM bases on propositional temporal logic (PTL) which is considered expressive enough for high-level specifications. Messages are interpreted as predicates that hold in the state after the sending/receiving of the message.

In contrast to TROLL, all properties are strongly related to the behavior and are specified in PTL. TROLL aims at providing concepts that can be described intuitively, which results in more readable an easier to understand specifications.

CMSL

CMSL [Wie90, Wie91c] is a language for the specification of conceptual models. CMSL combines approaches to algebraic specification of data types, semantic data modeling, and algebraic specification of processes.

Like TROLL, CMSL distinguishes data values and objects: in contrast to values, objects have a state. Through an identifier or proper name, we are able to distinguish distinct objects being in the same state, and we may relate different states of the same object.

CMSL is an algebraic language for the specification of a model of a Universe of Discourse, which is regarded (as in our approach) as an object system. A model contains knowledge about the object system. A specification defines the structure and behavior of elements in the conceptual model.

CMSL Version 2 [Wie91c] has two sub-languages:

- The *Value Specification Language* VSL which is an algebraic language for the specification of abstract data types.

- The Object Specification Language OSL for the specification of abstract object types (AOTs). An object type is specified through

    - attributes, which are modeled as functions mapping an object identifier to a data value;

    - events being the update operations on attributes; and

    - life cycles or processes, which define the admitted sequences of events.

Events are modeled as functions where the parameter sorts are domains and a special event sort is the codomain. The effects of event occurrences on attribute values are specified by positional terms like in TROLL. Furthermore, events can be synchronized using an explicit synchronization operator and messages to define the local effects of (global) communication events.

The specification of the dynamic behavior of objects in CMSL is based on process algebra. Thus, the underlying process algebra can be specified within the algebraic framework. Life cycles of objects are specified as processes, i.e. similar to our pattern specifications (CMSL Version 2).

The import relations between value specifications, object specifications, and process specifications are illustrated by Figure 9.1 (from [Wie91c]).

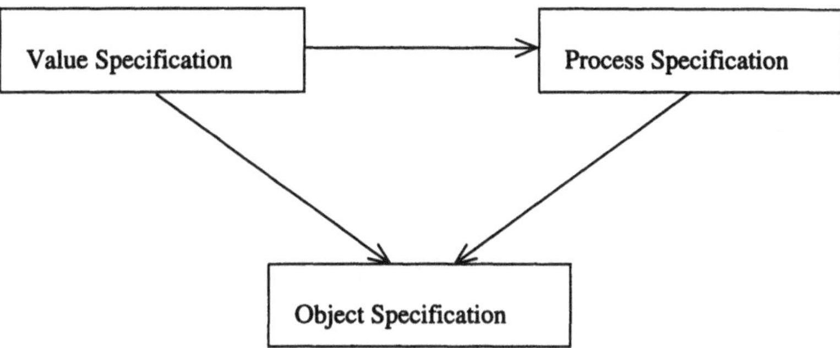

Figure 9.1: Import Relation between Specification Parts in CMSL

CMSL supports the concept of role besides static specialization. It has been one of the first proposals to formally define this concept.

CMSL is fully based on algebraic specification. The interpretation structures are *version algebras*: each possible state (in terms of a tuple of attribute values) is represented by an element of the algebra, and admissible state transitions are defined as functions between states. The algebra of all possible states and state transitions makes up the semantics of an object.

On the language side, there are a lot of similarities between TROLL and CMSL. CMSL, however, supports a much more algebraic style of specification. Therefore, many properties which are implicit in TROLL must be specified explicitly in CMSL, among them the structure of identifier spaces. As a drawback must be mentioned that Version 2 of CMSL does not support a declarative style of the specification of dynamic behavior in terms of permissions and obligations.

Glider

Glider (*G*eneral *L*anguage for the *I*ncremental *D*efinition and *E*laboration of *R*equirements) is a language which is developed in the ESPRIT-II project ICARUS [DDRW91, DDR92]. Its basic principles are:

- *Expressiveness*: Requirements analysis is not a coding task. Thus, a language must provide a number of abstraction mechanisms to allow for a natural mapping of user requirements to specifications. As such, the language supports the distinction of *objects* and *values*, of dynamic constraints to specify the evolution of objects over time, and the handling of increments (see Page 13). Since real-world systems are intrinsically time-varying, the concepts of *object*, *event*, and *time* itself are built in.

- *Structuring Mechanisms*: Large specifications must be organized into units with defined relationships between them.

- *Formality*: A specification language must support the rigorous interpretation of specifications in order to be precise, and to allow analysis and validation of specifications.

Glider is based on many-sorted first order logic with algebraic semantics. That is, the language basically follows the style described in Section 3.1, although other specification styles are integrated.

The language concepts in Glider are the following:

- *Type Clusters* are specifications of data types. A type may be constructed over other types or may be generated by generator functions. Furthermore, *operations* are defined over sorts. Operations are specified over a signature using preconditions and postconditions (*assertions*). For type clusters, explicit *provide window* may be defined—these are interfaces making available operations to the environment.

- *Objects* are entities of a particular type whose properties may vary over time. A *state* of an object is a data value that is structured according to the object's type. It can be accessed using an operation on the object with reference to a point in time. Assertions for objects are given in temporal logic, thus allowing for the declarative description of behavior.

- *Events* are (global) instantaneous occurrences that affect the state of several objects. Events are specified as type clusters also, thus allowing for the specification of preconditions and impacts on participating objects. Events in Glider are similar to relationships with global communication in TROLL.

- Glider uses global, linear time. There are not concrete references to single points in time, but in specifications we may compare points in time to obtain a concrete (real-time) *duration*.

- Glider supports (syntactical) inheritance of specifications. In contrast to TROLL, inheritance is not possible on the instance level: when a specification inherits from another, the defined sorts are distinct. On the specification level, only particular properties and operations may be inherited (interfacing).

- Glider supports *parameterized* types. These can be given different interpretations depending on the actual instantiation of the formal parameters.

On the conceptual level, Glider is closely related to TROLL. The language is, however, heavily influenced by algebraic specification. Its semantics is algebraic in nature. Glider does not support high-level abstraction mechanisms like specialization or roles. Composite objects can be specified by composite types only, thus the composition is static. Glider, however, has an implicit notion of time and thus supports the specification of real-time systems.

A more recent development based on Glider has been described in [DDR92]. It supports the top-down decomposition of composite systems into components. The system and the environment are separated and are related through interfaces.

ALBERT and Clyder

Based on Glider, the same group developed the language ALBERT (*A*gent-Oriented *L*anguage for *B*uilding and *E*licitating *R*equirements for Real-*T*ime Systems). ALBERT is a formal language that provides graphical features for structural descriptions and declarations and textual features for expressing various types of constraints. In parallel, the language Clyder was developed also on the basis of Glider [HRW93] with very similar concepts and motivations.

The main concepts behind ALBERT are those of *agent*, *action* and *perception*. An agent models an entity having *local contractual responsibilities*. An action models a discrete change occurring in a system. Perceptions allocated to agents model the knowledge an agent has about the behavior of other agents (its environment). An agent specification consists of a *declaration* part and a *constraints* part.

In the declaration part, the state structure of an agent is defined in terms of components of certain types and the possible actions are declared. The types of components can be (similar to attributes in TROLL) data types and object identifier types. Static visibility relationships between agents are explicitly specified in the declaration part by importing and exporting components from/to other agents. Agents in a system are hierarchically related through product and set constructors.

In the constraints part, the possible behavior, as well as the perceptions and informations are specified in terms of logical formulas of various kinds. *Basic* constraints describe the initial state of an agent. *Local* constraints specify the internal behavior of an agent. In the STATE BEHAVIOUR section we may specify dynamic constraints on the state evolution (similar to constraints in TROLL). The EFFECTS OF ACTIONS section specifies the local effects of action occurrences (as in TROLL). In the CAUSALITY section, causality relationships between actions can be stated. Formulas in the CAPABILITY section express obligations (an action has to occur in a situation), preventions (an action is not allowed to occur in a situation) and exclusive obligations (an action must occur under certain conditions and cannot occur otherwise). *Cooperation* constraints specify the interaction of a component with its environment. In the ACTION PERCEPTION section we may specify how external action occurrences are allowed to affect the local behavior of an agent. STATE PERCEPTION formulas control the perception of state information imported from other agents. In the ACTION INFORMATION and STATE INFORMATION sections we may specify which information about occurring actions and the current state is made available to other agents.

The language Clyder is a very similar approach differing mainly in the cooperation specification and the behavior specification. Clyder distinguishes agents and objects of which only the former may initiate changes and are capable of perceiving a limited extend of their environment. Perception of agents can only be specified statically, i.e. only static visibility relations between attributes and events of agents can be specified.

Central to the description of agent behavior are *reaction clauses* that specify interaction and temporal relationships between events as well as the effects of events on the state of agents. *Actions* are composed from events and are regarded to have a duration. Consequently, actions can be interrupted by other events—the effects of interrupts can be specified.

Along with ALBERT and Clyder, guidelines for the development of specifications from informal descriptions are provided.

Both languages are very similar to TROLL in that they model a system in terms of agents (or objects) that have an observable state and that evolve in a discrete manner. However, both languages incorporate real-time temporal logic which is not present in TROLL. Especially ALBERT puts emphasis on the local specification of the cooperation of an agent with others and thus seems to be better suited to describe systems at a very high level of abstraction.

Other Approaches

A number of other approaches with similar goals have been proposed in the literature. An overview can also be found in [RC92].

- TROLL*light* [CGH92, GCH93] is a dialect of TROLL especially tailored for verification purposes. Compared to TROLL, abstraction mechanisms have been thrown out and only a CSP-like sublanguage is offered for the specification of behavior. Systems are specified through a hierarchy of template specifications where templates are the only building blocks available in TROLL*light*.

- ERAE [DHR91] is a formal specification language for the conceptual modeling of dynamic systems. It is basically a many-sorted first-order real-time temporal logic and supports the declarative specification of systems composed from entities. Furthermore, ERAE provides the notion of event which is modeled as a special sort denoting (global) occurrences of events. It provides some structuring mechanisms which, however, do not rely on the notion of object.

- TEMPORA [TWBL90] is another approach putting emphasis on the modeling of temporal behavior. In TEMPORA, a conceptual model has three components: a structural model (called Entity-Relationship-Time model), a Rule Model and a Process Model. The structural model is an ER-model (ERT) extended with concepts to describe composite objects and specialization. In addition, concepts to timestamp constructs are offered. The Process Model is based on dataflow diagrams. The Rule Language is used for the declarative specification of constraints, derivations, and behavior and may constrain both the Process Model and the ERT model. The Rule Language is based on first-order temporal logic.

- Other conceptual modeling approaches that integrate the modeling of behavior are the ones described in [Lip90] and [Saa91]. These approaches use temporal logic for the separate description of dynamic aspects (see also Section 3.2).

- Languages like TAXIS [MBW80] or RML [GBM86] support the separate description of entities along with relationships between them and transactions or activities, which define scripts with several participating objects.

- The approaches [RVS91] and MADIS [EE91] use global events for reconciling the structural and behavioral view of a system. The languages proposed by the authors are quite similar to TROLL. Compared to TROLL, however, the approaches support the separate specification of objects, events, and rules which results in less support of localization of properties. O* [Bru91] is another approach being very similar to TROLL.

- DisCo (*Dis*tributed *Co*operation) [Kur91, KSV91, Jär92]is a language supporting the specification of distributed, reactive systems at a high level of abstraction. In DisCo, actions are not local to objects but shared between participating objects—thus, there are two fundamental concepts: *objects* and *actions*. DisCo supports modularity and transformation of specifications. It is based on an interleaving model. Compared to TROLL, DisCo does not support the integration of local state and local state transitions.

- Other rigorous object-oriented languages for conceptual modeling are TRIO+ [MS91], and ABACUS [NP90]. These languages put emphasis on the specification of temporal behavior either using temporal logic (TRIO+) or process specifications (ABACUS).

- FOOPS [GM87] is a formal language for object-oriented specification. It is based on the algebraic specification language OBJ2 which is included as a sub-language. In FOOPS, objects and values are distinguished. Values are data values in the sense of functional programming or algebraic specification, whereas objects have a state which can be altered by methods. The identification spaces in TROLL correspond to classes in FOOPS which are types of instance identifiers. Attributes are notated as functions associating values with object identifiers. Methods are notated as functions that transform a state into another state. The semantics of methods is specified by pre- and postconditions. Object-valued attributes are used to construct composite objects. FOOPS also has declarative language constructs for error handling which is missing in TROLL. Furthermore, we may specify parameterized classes.

  FOOPS does not support the specification of object behavior over time and interaction, although it not generally prohibits it.

- Maude [Mes90] is related to FOOPS in that it also incorporates OBJ2. The dynamics, however, is treated much more explicit than in FOOPS. Maude is based on Concurrent Rewriting (see Page 45).

9.2   Graphical Object-Oriented Formalisms

In this section we are briefly presenting some graphical formalisms of which the concepts are related. One characteristic of all approaches is that not everything can be specified in only one diagram—several types of diagrams are needed to specify different aspects of the system.

Graphical OBLOG

Graphical OBLOG [SRGS92] is based on the languages OBLOG and OBL-89. A system is modeled as a collection of objects, an object community. Objects are organized in object classes which provide identification spaces (called *surrogate spaces* there). The signature of an object (its attributes and events) is called a *matrix*. A system is specified using different diagrams:

- In the *Surrogate/Inheritance Diagram*, the classes along with its surrogate spaces as well as relationships between classes are specified. Surrogates are tuples of data values, where the projection functions are interpreted as key attributes.

  Surrogate spaces can be constructed over others as products (tuples), replication, or duplication.

  Inheritance may be described by subsort relationships on the surrogate level called *selections* (resulting in subclass relationships on the instance level).

- The *Matrix Diagram* defines the local signature of objects. The diagram depicts the key attributes and state dependent attributes with their codomains, and the events along with their arity. Attributes can be marked by static constraints restricting the possible values.

- In a *Behavior Diagram*, the admissible life cycles of objects are defined as sequences of events. The behavior is described as state machines, where the events cause transitions between *situations*. A life cycle always must start with a birth event and may end with a death event. Events can be equipped with a precondition that must be fulfilled to permit the event to occur.

- In *Attribute Initialization/Updating Diagrams*, the effects of event occurrences on attribute values are specified.

- *Interaction Diagrams* are used to specify interactions between objects in terms of event calling (see Section 7.2).

- *Value Binding Diagrams* are used to specify the parameters of events in the presence of data flow. These diagrams are used to represent special conditions in Behavior Diagrams, Attribute Updating Diagrams, and Interaction Diagrams.

In comparison to TROLL, Graphical OBLOG allows to specify behavior only by state machines. This can sometimes become difficult to read when an object can be in a large number of situations. Behavior diagrams cannot be nested or split and thus may suffer from state space explosions.

Since Graphical OBLOG and TROLL have the same ancestor, OBLOG, there are a lot of similarities between the concepts. The language is now supported by a tool [ESD93].

Object/Behavior Diagrams

Object/Behavior Diagrams [KS90, KS91] use a notion based on an extended Entity-Relationship model for the structural part and Petri nets for the behavioral part. A system specification consists of a number of diagrams that cover different aspects of the system to be described:

- *Object Diagrams* describe the structural aspects of systems. The authors use a ER-based notation. An object type is described by attributes (single-valued or set-valued) and by relationships to other object types. The model supports a number of relationship types:

  - A *general relationship* is a directed relationship between independent objects.

  - A *has-component* relationship models relationships between composite objects and dependent components.

  - A *has-constituent* relationship describes an aggregation between independent objects.

  - A *role-of* relationship defines aspects of the same conceptual object.

- *Life Cycle Diagrams* define state of objects, state transitions, and possible sequences of state transitions in terms of a Petri net based notation (see Page 43). States of an instance are explicitly named and correspond to places in a Petri net, events (or activities) correspond to transitions and describe state transitions.

  On the level of object life cycles, general synchronization constraints may be expressed.

  - *Domain restriction constraints* are dependencies that must be fulfilled by relationship properties in subsequent states, i.e. they restrict possible relationships to other objects.

  - *Hierarchical synchronization constraints* synchronize the life cycles of objects with the life cycles of their components.

  - *Triggered synchronization constraints* describe that certain activities are triggered by activities local to other objects (suffered events).

- In *Activity Specification Diagrams* the events or activities are specified in terms of their signature and by pre- and postconditions. Please note that in activity specification diagrams the events or activities may be shared by several objects as in Petri nets. Activities have input parameters and output parameters and must fulfill preconditions in order to be executed and must fulfill postconditions after execution.

- In *Activity Realization Diagrams* the implementation of abstract activities is specified. Only an activity which is local to an object may change local properties. The implementation is specified using Nassi-Shneiderman diagrams.

- *Activity Invocation Diagrams* describe communications between objects in terms of message passing. Messages are input and output parameters for activities and correspond to tokens in Petri nets and thus are similar to event calls.

Object/Behavior Diagrams support a very intuitive style of specification. The concepts behind Object/Behavior Diagrams are similar to the ones behind TROLL. The specification, however, follows a more operational style using Petri net ideas. Declarative features like pre- and postconditions of activities are mixed up with operational aspects like the implementation of activities.

Object Modeling Technique

OMT [RBP$^+$91] is a proposal for the object-oriented modeling of software systems. A system description is given by three kinds of models that are described by diagrams and that capture different aspects of the system to be modeled: the *functional model* describes what happens in the entire system, the *dynamic model* describes how and when it happens and the *object model* describes the object structures to which it happens.

- The *Object Model* is described by a diagram that is similar to an extended ER-Diagram. It shows the object classes in a system and the relationships between them. Objects have attributes and are identified by implicit unique identifiers. Furthermore, objects have associated operations, of which the signature is defined in the object model. In the object model, we may define *relationships* between classes in the sense of relationships in the ER-Model. An instance of a relationship is called a link.

  The object model supports the abstraction mechanisms aggregation and generalization. Generalization implies inheritance, i.e. the sharing of properties. Generalization implies also an is-a relationship between instances, i.e. we have specialization also on the instance level.

  Additionally, we may define keys, and certain static constraints in the object model.

- In the *Dynamic Model*, extended finite state machines are used to describe the behavior of objects. The notation follows the Statechart notation [Har88]. A state is an abstraction of attribute values and links of an object. Events are named, instantaneous one-way communications which, sometimes with parameters. On receiving an event, a state machine may change state, may send other events, or perform an action. An action is an update on attribute values or links of an object. An object may also react on entering a state by executing an operation. In contrast to events and actions, operations take time. State diagrams may be structured by nesting.

- In the *Functional Model* the behavior of the *system* rather than the behavior of an object is described. OMT uses data flow diagrams (DFDs) for this purpose, showing the data flow from sources through processes to destinations. The primary purpose is to describe how output values are derived from input values of processes by relating processes to operations on objects.

OMT puts emphasis on the description of the structural aspects. It stands in the tradition of semantic data modeling.

One big problem with OMT is that the functional model is not object-oriented but global. Thus, it is difficult to integrate the functional model with the other models where properties are grouped around objects.

A second problem concerns the entire description of the dynamic aspects: the differences between the notions of event, action, and state are not clear enough, and there does not exist a precise definition of inheritance of behavior. Furthermore, the issue of interaction between objects cannot be specified in the dynamic model. It has been stated in [HC91] that the three models are not coherent. OMT is, however, appealing due to the comprehensible notation compared to a number of approaches in that field.

Other Approaches

Other approaches are mostly even less formal than OMT and thus are not precise enough for a rigorous specification of requirements. A characteristic feature of these approaches is that they rely on traditional notations and thus mix up functional decomposition and semantic modeling. Comparisons are given in [CF92] and [MP92b].

- OOA by Coad and Yourdon [CY89] basically relies on entity-relationship modeling and object-oriented programming concepts. The emphasis lies on the structural aspects—the description of behavioral properties is not very much supported. The semantics of the modeling features are not precisely defined.

- Object Lifecycle Analysis by Shlaer and Mellor [SM92] besides features for structural modeling provides also features for the modeling of object dynamics and for modeling communication relationships between objects. Again, the approach is not defined in a very precise way and especially the coherence between the different diagrams is missing.

- OOD by Booch [Boo90] is an approach to support the high-level object-oriented design of software systems. The approach provides a large number of features which are sometimes confusing. This is mainly due to the lack of precision in defining the concepts.

- A recent approach is Objectory [Jac92] of which the basic concept is that of a *use-case*. A use-case is an admissible sequence of communications between objects.

10 Discussion and Outlook

10.1 Summary

In this thesis, we have presented the language TROLL for the logic-based specification of dynamic object systems. Dynamic object systems result from viewing problem domains or Universes of Discourse through an *object-oriented perspective*: a problem domain consists of components or objects that evolve concurrently in a discrete, event-driven manner.

The language is particularly suited to be used in the early phases of the development of information systems where existing systems have to be described or systems to be developed have to be prescribed (requirements specification or conceptual modeling phase).

The design of the language has been guided by the following development principles for large software systems:

- *Formality*: A language should be formally founded. Only then a specification allows for a rigorous, precise interpretation. Furthermore, we may derive further knowledge from a specification using theories generated from specifications.

- *Expressiveness*: A language to be used in the early phases of system design must allow for the 'natural' representation of user requirements without forcing specifiers to code.

- *Structuredness*: A language should support the organization of the specification in terms of (perhaps related) units. The relationships between units must be formally defined.

The basic characteristics of objects in our approach are the following:

- Objects have observable properties (*attributes*) that can change their values over time.

- State changes are solely possible through the occurrence of local *events*.

- Objects may interact through event calling.

- We may model roles of objects, specializations of objects and composite objects.

- Independent objects may be related by relationships describing communications or dependencies.

The language TROLL which is defined in the main part of this thesis supports the abstract description of dynamic object systems.

- The basic structuring mechanism of the language is an object specification, called a *template*. A template comprises the description of a local signature (typed attributes and local events) and the description of admissible observations, state changes, and behavior of an instance described by the template. The behavior may be described in several ways using declarative permissions and obligations as well as operational descriptions of event patterns.

- The language provides a number of *abstraction mechanisms*:

  - Specialization as a more detailed view of a concept (*is-a relationship*);

  - Roles as a temporary dynamic specialization; and

  - Composite objects as objects being constructed from others (*part-of relationship*).

- The language also provides constructs that support putting together components of a system. The constructs also allow the extension of systems and some modularization above the object level. The constructs are:

  - Relationships defining global constraints, global interactions and scripts;

  - Interfaces provide restricted views on separately defined units in order to hide unnecessary details to other system components.

The semantics of the language is defined using the *Object Specification Logic* [SSC92]. Basically, the logic supports the specification of temporal behavior by built-in predicates for enabledness and occurrence of events and and temporal connectives. The logic allows for the specification of concurrent behavior. Furthermore, it has an inherent notion of locality. The advantages of a logical approach are

- the possibility of incremental specification, i.e. a specification can be refined or completed by adding specifications of additional properties, and

- we may derive further knowledge about properties of the system from a specification using the inference rules of the logic.

10.2 Discussion

There have been many discussions where the number of language features have been questioned. We are strongly in favor of providing a rich variety of language features in order to support the high-level specification of problem domains and systems. As said above, a specifier in this phase shall not be forced to encode his understanding of system components using a small number of features since in this phase not coding is the task but understanding and representing requirements.

Concerning the semantics of certain constructs in TROLL, we can identify some decisions that can be questioned:

- The semantics of roles does not directly reflect the fact that we are talking about aspects of the same object since we model roles as aggregations of a base part and a role part. We made that decision in order to be able to distinguish several plays of the same role by one base aspect. An alternative way of modeling would be to require that additional properties are not valid if a role is not played and that inherent properties model the differences between same roles played at different times.

- The OSL representation of complex objects is very similar to that of roles. A five-level construction must be questioned. The reason for pursuing the construction is the OSL only supports static template morphisms. Thus, the modeling of dynamic compositions required a multilevel construction. This is clearly a point where some additional work must be done.

For interfaces in TROLL, the underlying logic OSL was too restrictive since the frame assumption is integrated into the logic. This rule prevents spontaneous changes in states (without events occurring) to happen. It has been argued in [ES91] that a semantical framework must not insist on a frame assumption in order to allow abstractions. In this thesis, we stuck to OSL since the goal was to define the language TROLL. It should be examined if the restrictions in OSL can be dropped in order to obtain a more general framework. Then the frame rule can be introduced by the translation of TROLL into the logic which makes also possible the explicit specification of concrete frame rules connecting events and attributes more closely.

10.3   Further Work

A language hardly is ready at some point in time. Like requirements, there are always decisions that have been made on the basis of assumptions only. The language features have to be examined now. We can at this moment identify problems that have to be solved in the near future:

- The semantics of roles and composite objects in OSL have to be further discussed.

- The modeling of activity is not clear and needs further work.

- A notion of transaction as process that is guaranteed to be executed or not at all should be integrated.

- The description of activities of long duration with several participating objects (scripts) must be better supported. Especially in the analysis phase the need to model business procedures and processes has been realized.

- The issue of higher levels of modularization needs further work.

- Interaction should not be handled in a very strict way but with the possibility of handling exceptions when an interaction cannot be carried out as specified. A diploma thesis is currently examining possible ways to integrate exception handling into the language [Sch92a].

A parser supporting a simple analysis of specifications for the precursor version TROLL91 [JSHS91] has been implemented by two students [Ste92, Stö92].

A textual language is quite difficult to use, especially when specifications grow in size. Therefore, a graphical front-end is desperately needed. As we have seen in Section 9.2, it is not always guaranteed that several diagrams describing different aspects of a system work together well. TROLL could provide a basis that supports the analysis of several diagrams and the guidance through diagrams showing different properties of objects.

Of course there are many issues that have not been discussed in this thesis but are essential for information systems.

- Assistance is needed in design concerning *libraries* and reusability of specification units. The assistance must address retrieval, adaption, and management of reusable components. TROLL itself provides a high level of structuring and reuse capabilities, but the whole issue needs further discussions.

- Query interfaces [JSS90, SJ90] are essential for usage of systems specified in TROLL. At the conceptual level, however, we do not have to worry about the operational semantics of queries but we have to be able to declaratively specify queries over systems.

- Specifications of systems must be implemented, otherwise they are useless. We can identify two orthogonal directions here [HJSE92]:

  - *Animation* means to execute specifications on the same level of abstraction. This means that objects as they are specified are translated into executable programs or specifications. Animation aims at providing a way to let the user evaluate a specification and not the implementation of a specification in the target environment. A possible framework for executable specifications describing a distributed system of interacting objects has been published in [HJ91].

  - *Implementation* is the realization of the system as specified in a target environment. Here, we have to transform core components into computerized representations whereas other components have to be transformed into interfaces to system components that exist physically in the environment of the information system. In contrast to animation, implementation cannot be carried out without design decisions made by humans. Implementations have, however, to be verified against specifications. It is still an open issue how this can be done.

- Today, many applications have to take legacy systems into account and have to integrate such systems [Bro92]. Our approach has been applied to describe global applications on top of existing systems [SJH93] in a prototypical way. Here, a lot of work has still to be done, e.g. translation of specifications to interfaces of existing systems, and the evaluation of the language in this application area through case studies.

- In order to work with specifications, information about the specification has to be available. In traditional terms, a *data dictionary* is responsible for this task. Ideally, a dictionary should be modeled in the same formalism as the model itself, i.e. we need to do *meta-modeling* of the specification language. This issue has just been put on the shopping list for further work and no results are available.

This thesis has just contributed a single brick to a framework of interactive information systems development. This task involves many other problems concerning e.g. support of cooperative work, version support, animation, implementation, and documentation (just to mention a few research problems).

Finally, a language like TROLL must be backed by *guidelines* for using it. In analysis, it seems not suitable to force analysts to follow a certain method, but a collection of guidelines supporting the elicitation and the specification of a conceptual model is needed.

Bibliography

[ABD+89]   Atkinson, M.; Bancilhon, F.; DeWitt, D.; Dittrich, K. R.; Maier, D.; Zdonik, S. B.: The Object-Oriented Database System Manifesto. In: Kim, W.; Nicolas, J.-M.; Nishio, S. (eds.): *Proc. Int. Conf. on Deductive and Object-Oriented Database Systems*, Kyoto, Japan, December 1989. pp. 40–57.

[Abi90]   Abiteboul, S.: Towards a Deductive Object-Oriented Database Language. *Data & Knowledge Engineering*, Vol. 5, No. 2, 1990, pp. 263–287.

[Agh86]   Agha, G. A.: *ACTORS: A Model of Concurrent Computation in Distributed Systems*. The MIT Press, Cambridge, MA, 1986.

[AH87]   Abiteboul, S.; Hull, R.: IFO—A Formal Semantic Database Model. *ACM Transactions on Database Systems*, Vol. 12, No. 4, 1987, pp. 525–565.

[AHT90]   Allen, J.; Hendler, J.; Tate, A. (eds.): *Readings in Planning*. Morgan Kaufmann, San Mateo, CA, 1990.

[Ara91]   Arapis, C.: Temporal Specifications of Object Behavior. In: Thalheim, B.; Demetrovics, J.; Gerhardt, H.-D. (eds.): *Proceedings 3rd. Symp. on Mathematical Fundamentals of Database and Knowledge Base Systems MFDBS–91*, Rostock, 1991. LNCS 495, Springer-Verlag, Berlin, 1991, pp. 308–324.

[Ara92]   Arapis, C.: *Dynamic Evolution of Object Behavior and Object Cooperation*. PhD thesis, Université de Genève, Geneva, 1992.

[BB84]   Batory, B.; Buchmann, A.: Molecular Objects, Abstract Data Types and Data Models: A Framework. In: Dayal, U.; Schlageter, G.; Seng, L. H. (eds.): *Proc. 10th Int. Conf. on Very Large Databases VLDB'84*, Singapore, 1984. pp. 172–184.

[BDMN73] Birtwistle, G.; Dahl, O.; Myhrtag, B.; Nygaard, K.: *Simula Begin*. Auerbach Press, Philadelphia, 1973.

[Bee90] Beeri, C.: A Formal Approach to Object Oriented Databases. *Data & Knowledge Engineering*, Vol. 5, No. 4, 1990, pp. 353–382.

[Boo90] Booch, G.: *Object-Oriented Design*. Benjamin/Cummings, Menlo Park, CA, 1990.

[Bor85] Borgida, A.: Features of Languages for the Development of Information Systems at the Conceptual Level. *IEEE Software*, Vol. 2, No. 1, 1985, pp. 63–73.

[Bra92] Braß, S.: *Defaults in deduktiven Datenbanken*. PhD thesis, Universität Hannover, 1992. *In German*.

[Bro92] Brodie, M. L.: The Promise of Distributed Computing and the Challenges of Legacy Systems. In: Gray, P. M. D.; Lucas, R. J. (eds.): *Advanced Database Systems (Proc. 10th British National Conference on Databases BNCOD 10)*, Aberdeen (Scotland), 1992. LNCS 618, Springer-Verlag, Berlin, 1992, pp. 1–28.

[Bru91] Brunet, J.: Modeling the World with Semantic Objects. In: Van Assche, F.; Moulin, B.; Rolland, C. (eds.): *The Object-Oriented Approach in Information Systems*, Québec, Canada, 1991. North-Holland, Amsterdam, 1991, pp. 361–379.

[CF92] Champeaux, D. de; Faure, P.: A Comparative Study of Object-Oriented Analysis Methods. *Journal of Object-Oriented Programming*, Vol. 3, No. 2, 1992, pp. 21–33.

[CG92] Conrad, S.; Gogolla, M.: An Annotated Bibliography on Object-Orientation and Deduction. *ACM SIGMOD Records*, Vol. 21, No. 1, 1992, pp. 123–132.

[CGH92] Conrad, S.; Gogolla, M.; Herzig, R.: TROLL *light*: A Core Language for Specifying Objects. Informatik-Bericht 92–02, TU Braunschweig, 1992.

[Che76] Chen, P. P.: The Entity-Relationship Model—Toward a Unified View of Data. *ACM Transactions on Database Systems*, Vol. 1, No. 1, 1976, pp. 9–36.

[Cod79]   Codd, E. F.: Extending the Relational Model to Capture More Meaning. *ACM Transactions on Database Systems*, Vol. 4, No. 4, 1979, pp. 397–434.

[CSS89]   Costa, J.-F.; Sernadas, A.; Sernadas, C.: OBL–89 User's Manual (Version 2.3). Internal report, INESC, Lisbon, 1989.

[CY89]   Coad, P.; Yourdon, E.:   *Object-Oriented Analysis.*   Yourdon Press/Prentice Hall, Englewood Cliffs, NJ, 1989.

[DD92]   Dubois, E.; Du Bois, P.:   Reasoning on the Elaboration of Formal Requirements for Composite Systems. Internal report, Facultés Universitaires de Namur, Namur (B), June 1992.

[DDP93]   Dubois, E.; Du Bois, P.; Petit, M.:   O-O Requirements Analysis: An Agent Perspective. In: Nierstrasz, O. (ed.): *ECOOP'93—Object-Oriented Programming (Proc. 7th European Conference)*, Kaiserslautern, 1993. LNCS 707, Springer-Verlag, Berlin, 1993, pp. 458–481.

[DDR92]   Dubois, E.; Du Bois, P.; Rifaut, A.: Elaborating, Structuring and Expressing Formal Requirements of Composite Systems. In: Loucopoulos, P. (ed.): *Advanced Information Systems Engineering CAISE'92 (Proc. 4th Conf.)*, Manchester (UK), 1992. LNCS 593, Springer-Verlag, Berlin, 1992.

[DDRW91] Dubois, E.; Du Bois, P.; Rifaut, A.; Wodan, P.:   GLIDER Manual. ICARUS Deliverable, Facultés Universitaires de Namur, Namur (B), 1991.

[DeM79]   DeMarco, T.: *Structured Analysis and System Specification.* Prentice-Hall, Englewood Cliffs, NJ, 1979.

[DHR91]   Dubois, E.; Hagelstein, J.; Rifaut, A.:   A Formal Language for the Requirements Engineering of Computer Systems. In: Thayse, A. (ed.): *From Natural Language Processing to Logic for Expert Systems.* J. Wiley & Sons, Chicester, 1991, pp. 269–345.

[EDS93]   Ehrich, H.-D.; Denker, G.; Sernadas, A.: Constructing Systems as Object Communities. In: Gaudel, M.-C.; Jouannaud, J.-P. (eds.): *Proc. TAPSOFT'93: Theory and Practice of Software Development*, Paris, 1993. LNCS 668, Springer, Berlin, pp. 453–467.

[EE91]   Essink, L. J. B.; Erhard, W. J.: Object Modelling and System Dynamics in the Conceptualization Stages of Information Systems Development. In: Van Assche, F.; Moulin, B.; Rolland, C. (eds.): *The Object-Oriented Approach in Information Systems*, Québec, Canada, 1991. North-Holland, Amsterdam, 1991, pp. 89–116.

[EGH⁺92]   Engels, G.; Gogolla, M.; Hohenstein, U.; Hülsmann, K.; Löhr-Richter, P.; Saake, G.; Ehrich, H.-D.: Conceptual Modelling of Database Applications using an Extended ER Model. *Data & Knowledge Engineering*, Vol. 9, No. 2, 1992, pp. 157–204.

[EGL89]   Ehrich, H.-D.; Gogolla, M.; Lipeck, U. W.: *Algebraische Spezifikation abstrakter Datentypen*. Teubner, Stuttgart, 1989. *In German*.

[EGS90]   Ehrich, H.-D.; Goguen, J. A.; Sernadas, A.: A Categorial Theory of Objects as Observed Processes. In: Bakker, J. de; Roever, W. de; Rozenberg, G. (eds.): *Foundations of Object-Oriented Languages (Proc. REX School/Workshop)*, Noordwijkerhood (NL), 1990. LNCS 489, Springer-Verlag, Berlin, 1991, pp. 203–228.

[EGS91]   Ehrich, H.-D.; Gogolla, M.; Sernadas, A.: Objects and their Specification. In: Bidoit, M. et al. (eds.): *Recent Trends in Data Type Specification : Selected Papers (Proc. 8th Workshop on Specification of Abstract Data Types)*, Dourdan (France), 1991. LNCS 655, Springer-Verlag, Berlin, 1993.

[EM85]   Ehrig, H.; Mahr, B.: *Fundamentals of Algebraic Specification I: Equations and Initial Semantics*. Springer-Verlag, Berlin, 1985.

[Eme90]   Emerson, E. A.: Temporal and Modal Logic. In: Leeuwen, J. van (ed.): *Formal Models and Semantics*. Elsevier Science Publishers B.V., 1990, pp. 995–1072.

[EMO92]   Ehrig, H.; Mahr, B.; Orejas, F.: Introduction to Algebraic Specification. (Part 2: From Classical View to Foundations of System Specifications). *The Computer Journal*, Vol. 35, No. 5, 1992, pp. 468–477.

[EN89]   Elmasri, R.; Navathe, S. B.: *Fundamentals of Database Systems*. Benjamin/Cummings Publ., Redwood City, CA, 1989.

[ES89]   Ehrich, H.-D.; Sernadas, A.: Algebraic Implementation of Objects over Objects. In: de Bakker, J. W.; de Roever, W.-P.; Rozenberg,

G. (eds.): *Proc. REX Workshop "Stepwise Refinement of Distributed Systems: Models, Formalisms, Correctness"*, Mook (NL), 1989. LNCS 430, Springer-Verlag, Berlin, 1991, pp. 239–266.

[ES91]   Ehrich, H.-D.; Sernadas, A.: Fundamental Object Concepts and Constructions. In: Saake, G.; Sernadas, A. (eds.): *Information Systems—Correctness and Reusability. (Workshop IS-CORE '91, ES-PRIT BRA WG 3023, Selected Papers)*, London, 1991. TU Braunschweig, Informatik-Bericht 91-03, 1991.

[ESD93]   OBLOG CASE V1.0—The User's Guide. Espirito Santo Data Informatica, Lisbon, 1993.

[ESS89]   Ehrich, H.-D.; Sernadas, A.; Sernadas, C.: Objects, Object Types, and Object Identification. In: Ehrig, H.; Herrlich, H.; Kreowski, H.-J.; Preuß, G. (eds.): *Categorical Methods in Computer Science.* LNCS 393, Springer-Verlag, Berlin, 1989, pp. 142–156.

[ESS92]   Ehrich, H.-D.; Saake, G.; Sernadas, A.: Concepts of Object-Orientation. In: Studer, R. (ed.): *Proc. of the 2nd Workshop of "Informationssysteme und Künstliche Intelligenz"*, Ulm (FRG), 1992. IFB 303, Springer-Verlag, Berlin, 1992, pp. 1–19.

[Fea87]   Feather, M. S.: Language Support for the Specification and Development of Composite Systems. *ACM Transactions on Programming Languages and Systems*, Vol. 9, No. 2, 1987, pp. 198–234.

[FGJM85] Futatsugi, K.; Goguen, J. A.; Jouannaud, J.-P.; Meseguer, J.: Principles of OBJ2. In: Reid, B. K. (ed.): *Proc. 12th Symp. on Principles of Programming Languages.* ACM, New York, 1985, pp. 52–66.

[FM90]   Fiadeiro, J.; Maibaum, T.: Describing, Structuring and Implementing Objects. In: Bakker, J. de; Roever, W. de; Rozenberg, G. (eds.): *Foundations of Object-Oriented Languages (Proc. REX School/Workshop)*, Noordwijkerhood (NL), 1990. LNCS 489, Springer-Verlag, Berlin, 1991, pp. 275–310.

[FM92]   Fiadeiro, J.; Maibaum, T.: Temporal Theories as Modularisation Units for Concurrent System Specification. *Formal Aspects of Computing*, Vol. 4, No. 3, 1992, pp. 239–272.

[Fra86]   Francez, N.: *Fairness.* Springer-Verlag, New York, 1986.

[FSMS90]   Fiadeiro, J.; Sernadas, C.; Maibaum, T.; Saake, G.: Proof-Theoretic Semantics of Object-Oriented Specification Constructs. In: Meersman, R.; Kent, W.; Khosla, S. (eds.): *Object-Oriented Databases: Analysis, Design and Construction (Proc. IFIP WG 2.6 Working Conference DS-4)*, Windermere (UK), 1990. North-Holland, Amsterdam, 1991, pp. 243–284.

[GBM86]   Greenspan, S.; Borgida, A. T.; Mylopoulos, J.: A Requirements Modelling Language and its Logic. In: Brodie, M. L.; Mylopoulos, J. (eds.): *On Knowledge Base Management Systems*. Springer-Verlag, Berlin, 1986, pp. 471–502.

[GCH93]   Gogolla, M.; Conrad, S.; Herzig, R.: Sketching Concepts and Computational Model of TROLL *light*. In: Miola, A. (ed.): *Proc. 3rd Int. Conf. Design and Implementation of Symbolic Computation Systems (DISCO'93)*. LNCS Series, Springer-Verlag, Berlin, 1993.

[GH91]   Gogolla, M.; Hohenstein, U.: Towards a Semantic View of an Extended Entity-Relationship Model. *ACM Transactions on Database Systems*, Vol. 16, 1991, pp. 369–416.

[GJM91]   Ghezzi, C.; Jazayeri, M.; Mandrioli, D.: *Fundamentals of Software Engineering*. Prentice-Hall, Englewood Cliffs, NJ, 1991.

[GKK+87]   Goguen, J. A.; Kirchner, C.; Kirchner, H.; Mégrelis, A.; Meseguer, T., J. Winkler: An Introduction to OBJ3. In: Jouannaud, J.-P.; Kaplan, S. (eds.): *Proc. Conf. on Conditional Term Rewriting*, Orsay (F), 1987. LNCS 308, Springer-Verlag, Berlin, 1988, pp. 258–263.

[GM87]   Goguen, J. A.; Meseguer, J.: Unifying Functional, Object-Oriented and Relational Programming with Logical Semantics. In: Shriver, B.; Wegner, P. (eds.): *Research Directions in Object-Oriented Programming*. MIT Press, 1987, pp. 417–477.

[Gog92]   Goguen, J. A.: The Dry and the Wet. In: Falkenberg, E.; Rolland, C.; El-Sayed, E. N. (eds.): *Information Systems Concepts: Improving the Understanding*, Alexandria, 1992. IFIP Transactions A, North-Holland, Amsterdam, 1992, pp. 1–17.

[Gri82]   Griethuysen, J. J. van: Concepts and Terminology for the Conceptual Schema and the Information Base. Report N695, ISO/TC97/SC5, 1982.

[Har88]   Harel, D.:   On Visual Formalisms. *Communications of the ACM*, Vol. 31, No. 5, 1988, pp. 514–530.

[HC91]   Hayes, F.; Coleman, D.: Coherent Models for Object-Oriented Analysis. In: *Proc. OOPSLA'91*, Phoenix, 1991. *Special Issue of SIGPLAN Notices, Vol. 26, No. 11, ACM, 1991*, pp. 171–183.

[Heu92]   Heuer, A.: *Objektorientierte Datenbanken.* Addison-Wesley, Bonn (D), 1992. *In German.*

[Hew77]   Hewitt, C.:   Viewing Control Structures as Patterns of Passing Messages. *Artificial Intelligence*, Vol. 8, 1977, pp. 323–364.

[HG88]   Hohenstein, U.; Gogolla, M.:   A Calculus for an Extended Entity-Relationship Model Incorporating Arbitrary Data Operations and Aggregate Functions. In: Batini, C. (ed.): *Proc. 7th Int. Conf. on the Entity-Relationship Approach*, Rome, 1988. North-Holland, Amsterdam, 1988, pp. 129–148.

[HJ91]   Hartmann, T.; Jungclaus, R.:   Abstract Description of Distributed Object Systems. In: Tokoro, M.; Nierstrasz, O.; Wegner, P. (eds.): *Object Based Concurrent Computing (Proc. ECOOP'91 Workshop)*, Geneva (CH), 1991. LNCS 612, Springer-Verlag, Berlin, 1992, pp. 227–244.

[HJS92]   Hartmann, T.; Jungclaus, R.; Saake, G.:   Aggregation in a Behavior Oriented Object Model. In: Lehrmann Madsen, O. (ed.): *ECOOP'92 European Conference on Object-Oriented Programming (Proceedings)*, Utrecht (NL), 1992. LNCS 615, Springer-Verlag, Berlin, 1992, pp. 57–77.

[HJSE92]   Hartmann, T.; Jungclaus, R.; Saake, G.; Ehrich, H.-D.: Spezifikation von Objektsystemen. In: Bayer, R.; Härder, T.; Lockemann, P. C. (eds.): *Objektbanken für Experten.* Springer, Berlin, Reihe Informatik aktuell, 1992. *In German.*

[HK87]   Hull, R.; King, R.: Semantic Database Modeling: Survey, Applications, and Research Issues. *ACM Computing Surveys*, Vol. 19, No. 3, 1987, pp. 201–260.

[HM81]   Hammer, M. M.; McLeod, D. J.: Database Description with SDM: A Semantic Database Model. *ACM Transactions on Database Systems*, Vol. 6, No. 3, 1981, pp. 351–381.

[HNSE87] Hohenstein, U.; Neugebauer, L.; Saake, G.; Ehrich, H.-D.: Three-Level Specification of Databases Using an Extended Entity-Relationship Model. In: Wagner, R.; Traunmüller, R.; Mayr, H. C. (eds.): *Proc. GI-Fachtagung "Informationsermittlung und -analyse für den Entwurf von Informationssystemen"*, Linz, 1987. Informatik-Fachbericht 143, Springer Verlag, Berlin, 1987, pp. 58–88.

[Hoa85] Hoare, C. A. R.: *Communicating Sequential Processes.* Prentice-Hall, Englewood Cliffs, NJ, 1985.

[Hoh90] Hohenstein, U.: *Ein Kalkül für ein erweitertes Entity-Relationship-Modell und seine Übersetzung in einen relationalen Kalkül.* PhD thesis, Technische Universität Braunschweig, 1990. *In German.*

[Hoh93] Hohenstein, U.: *Formale Semantik eines erweiterten Entity-Relationship-Modells.* Teubner-Verlag, Stuttgart/Leipzig, 1993. Revised version of [Hoh90], in German.

[HR92] Hagelstein, J.; Roelants, D.: Reconciling Operational and Declarative Specifications. In: Loucopoulos, P. (ed.): *Advanced Information Systems Engineering CAISE'92 (Proc. 4th Conf.)*, Manchester (UK), 1992. LNCS 593, Springer-Verlag, Berlin, 1992.

[HRW93] Hagelstein, J.; Roelants, D.; Wodon, P.: Formal Requirements Made Practical. In: *Proc. 4th European Software Engineering Conference ESEC'93*, Garmisch-Partenkirchen, 1993.

[HS93] Hartmann, T.; Saake, G.: Abstract Specification of Object Interaction. Internal Report, TU Braunschweig, 1993.

[ISO84] ISO: Information Processing Systems, Definition of the Temporal Ordering Specification Language LOTOS. Report N1987, ISO/TC97/16, 1984.

[Jac92] Jacobson, I.: *Object-Oriented Software Engineering.* Addison-Wesley, Reading, MA, 1992.

[Jär92] Järvinen, H.-M.: *The Design of a Specification Language for Reactive Systems.* PhD thesis, Tampere University of Technology, 1992.

[JHS92] Jungclaus, R.; Hartmann, T.; Saake, G.: Relationships between Dynamic Objects. In: Kangassalo, H. (ed.): *Information Modelling* and Knowledge Bases IV: Concepts, Methods and Systems* (Proc. 2nd Eurpean-Japanese Seminar on Information Modelling and Knowledge Bases, Ellivuori (SF), 1992. IOS Press, Amsterdam, 1993, pp. 425–438.

[JSH91]   Jungclaus, R.; Saake, G.; Hartmann, T.:   Language Features for Object-Oriented Conceptual Modeling. In: Teory, T. J. (ed.): *Proc. 10th Int. Conf. on the ER-approach*, San Mateo, 1991. pp. 309–324.

[JSHS91] Jungclaus, R.; Saake, G.; Hartmann, T.; Sernadas, C.:   Object-Oriented Specification of Information Systems: The TROLL Language. Informatik-Bericht 91-04, TU Braunschweig, 1991.

[JSS90]   Jungclaus, R.; Saake, G.; Sernadas, C.:   Using Active Objects for Query Processing. In: Meersman, R.; Kent, W.; Khosla, S. (eds.): *Object-Oriented Databases: Analysis, Design and Construction (Proc. 4th IFIP WG 2.6 Working Conference DS-4)*, Windermere (UK), 1990. North-Holland, Amsterdam, 1991, pp. 285–304.

[JSS91]   Jungclaus, R.; Saake, G.; Sernadas, C.: Formal Specification of Object Systems. In: Abramsky, S.; Maibaum, T. (eds.): *Proc. TAPSOFT'91 Vol. 2*, Brighton, 1991. LNCS 494, Springer-Verlag, Berlin, 1991, pp. 60–82.

[KBC⁺87] Kim, W.; Banerjee, J.; Chou, H.-T.; Garza, J. F.; Woelk, D.: Composite Object Support in an Object-Oriented Database System. In: Meyrowitz, N. (ed.): *OOPSLA'87 Conference Proceedings*. Special Issue of ACM SIGPLAN Notices, Vol. 22, No. 12, ACM, 1987, pp. 118–125.

[KC86]   Khoshafian, S. N.; Copeland, G. P.: Object Identity. In: *Proc. OOPSLA'86 Conference*, Portland, OR, 1986. Special Issue of SIGPLAN Notices, Vol. 21, No. 11, ACM, New York, 1986, pp. 406–416.

[Ken78]   Kent, W.: *Data and Reality*. North-Holland, Amsterdam, 1978.

[Ken91]   Kent, W.: A Rigorous Model of Object Reference, Identity, and Existence. *Journal of Object-Oriented Programming*, Vol. 4, No. 3, 1991, pp. 28–36.

[Kin89]   King, R.: My Cat Is Object-Oriented. In: Kim, W.; Lochovsky, F. H. (eds.): *Object-Oriented Concepts, Databases, and Applications*. ACM Press/Addison-Wesley, New York, NY/Reading, MA, 1989, pp. 23–30.

[KLS92]   Kramer, M.; Lausen, G.; Saake, G.: Updates in a Rule-Based Language for Objects. In: Yuan, Li-Yan (ed.): *Proc. 18th Int. Conf. on Very Large Databases VLDB'92*, Vancouver, 1992. pp. 251–262.

[Kow89]   Kowalski, R.: The Limitations of Logic and Its Role in Artificial Intelligence. In: Schmidt, J. W.; Thanos, C. (eds.): *Foundations of Knowledge Base Management*. Springer-Verlag, Berlin, 1989, pp. 477–489.

[KS90]   Kappel, G.; Schrefl, M.: Using an Object-Oriented Diagram Technique for the Design of Information Systems. In: Sol, H. G.; van Hee, K. M. (eds.): *Dynamic Modelling of Information Systems (Proc. Int. Working Conf.)*, Noordwijkerhood, 1990. North-Holland, Amsterdam, 1991, pp. 121–164.

[KS91]   Kappel, G.; Schrefl, M.: Object/Behavior Diagrams. In: *Proc. Int. Conf. on Data Engineering*, Kobe, 1991. IEEE Computer Society Press, 1991, pp. 530–539.

[KS92]   Kappel, G.; Schrefl, M.: Local Referential Integrity. In: Pernul, G.; Tjoa, A. M. (eds.): *Entity-Relationship Aproach—ER'92 (Proc. of the 11th International Conference)*, Karlsruhe, 1992. LNCS 645, Springer-Verlag, Berlin, 1992, pp. 41–61.

[KSV91]   Kurki-Suonio, R.; Systä, K.; Vain, J.: Real-Time Specification and Modeling with Joint Actions. In: *Proc. 6th Int. Workshop on Software Specification and Design*, Como (I), 1991. ACM Press, New York, 1991.

[Kur91]   Kurki-Suonio, R.: Modular Modeling of Temporal Behaviors. In: Ohsuga, S. et al. (eds.): *Information Modelling and Knowledge Bases III*, Kyoto, Japan, 1991. IOS Press, Amsterdam, 1992, pp. 283–300.

[Lin90]   Lindgreen, P.: A Framework of Information Systems Concepts. Interim report, The IFIP WG 8.1 Task Group Frisco, Dept. of Information Systems, University of Nijmegen, 1990.

[Lip90]   Lipeck, U. W.: Transformation of Dynamic Integrity Constraints into Transaction Specifications. *Theoretical Computer Science*, Vol. 76, 1990, pp. 115–142.

[LL90]   Lamport, L.; Lynch, N.: Distributed Computing: Models and Methods. In: Leeuwen, J. van (ed.): *Formal Models and Semantics*. Elsevier Science Publishers B.V., 1990, pp. 1157–1199.

[MBJK92] Mylopoulos, J.; Borgida, A.; Jarke, M.; Koubarakis, M.: Telos: Representing Knowledge About Information Systems. *ACM Transactions on Information Systems*, Vol. 8, No. 4, 1992, pp. 325–362.

[MBW80] Mylopoulos, J.; Bernstein, P. A.; Wong, H. K. T.: A Language Facility for Designing Interactive Database-Intensive Applications. *ACM Transactions on Database Systems*, Vol. 5, No. 2, 1980, pp. 185–207.

[Mes90] Meseguer, J.: A Logical Theory of Concurrent Objects. In: Meyrowitz, N. (ed.): *ECOOP/OOPSLA'90 Proceedings*, Ottawa, 1990. Special Issue of SIGPLAN Notices, Vol. 25, No. 10, ACM, New York, 1990, pp. 101–115.

[Mes92] Meseguer, J.: Conditional Rewriting Logic as a Unified Model of Concurrency. *Theoretical Computer Science*, Vol. 96, No. 1, 1992, pp. 73–155.

[Mes93] Meseguer, J.: Solving the Inheritance Anomaly in Concurrent Object-Oriented Programming. In: Nierstasz, O. (ed.): *ECOOP'93—Object-Oriented Programming (Proc. 7th European Conf.)*, Kaiserslautern, 1993. LNCS 707, Springer-Verlag, Berlin, 1993, pp. 220–246.

[Mil80] Milner, R.: *A Calculus of Communicating Systems*. Springer-Verlag, Berlin, 1980.

[Mil89] Milner, R.: *Communication and Concurrency*. Prentice-Hall, Englewood Cliffs, 1989.

[Mil90] Milner, R.: Operational and Algebraic Semantics of Concurrent Processes. In: Leeuwen, J. van (ed.): *Formal Models and Semantics*. Elsevier Science Publishers B.V., 1990, pp. 1201–1242.

[Min75] Minsky, M.: A Framework for Representing Knowledge. In: Winston, P. H. (ed.): *The Psychology of Computer Vision*. McGraw-Hill, New York, 1975, pp. 211–277. Reprinted in: Brachmann, R.; Levesque, H. (eds.): *Readings in Knowledge Representation*, Morgan Kaufmann, Los Altos, 1985.

[MP92a] Manna, Z.; Pnueli, A.: *The Temporal Logic of Reactive and Concurrent Systems —Vol. 1: Specification*. Springer-Verlag, New York, 1992.

[MP92b]   Monarchi, D. E.; Puhr, G. I.: A Research Topology for Object-Oriented
           Analysis and Design. *Communications of the ACM*, Vol. 35, No. 9,
           1992, pp. 35–47.

[MS91]   Morzenti, A.; San Pietro, P.: An Object-Oriented Logic Language
           for Modular System Specification. In: America, P. (ed.): *ECOOP'91
           European Conference on Object-Oriented Programming*, Geneva, 1991.
           LNCS 512, Springer-Verlag, Berlin, 1991, pp. 39–58.

[Myl90]   Mylopoulos, J.: Object-Orientation and Knowledge Representation.
           In: Meersman, R.; Kent, W.; Khosla, S. (eds.): *Object-Oriented
           Databases: Analysis, Design and Construction (Proc. IFIP WG 2.6
           Working Conference DS-4)*, Windermere (UK), 1990. North-Holland,
           Amsterdam, 1991, pp. 23–37.

[NP90]   Nierstrasz, O. M.; Papathomas, M.:   Viewing Objects as Pat-
           terns of Communicating Agents.   In: Meyrowitz, N. (ed.):
           ECOOP/OOPSLA'90 Proceedings, Ottawa, 1990. Special Issue of SIG-
           PLAN Notices, Vol. 25, No. 10, ACM, New York, 1990, pp. 38–43.

[Par90]   Partsch, H. A.: *Specification and Transformation of Programs: A
           Formal Approach to Software Development.* Springer-Verlag, Berlin,
           1990.

[Per90]   Pernici, B.: Objects with Roles. In: *Proc. ACM/IEEE Int. Conf.
           on Office Information Systems*, Boston, 1990. Special Issue of SIGOIS
           Bulletin, Vol. 11, No. 2&3, ACM Press, New York, 1990, pp. 205–215.

[PM88]   Peckham, J.; Maryanski, F.: Semantic Data Models. *ACM Computing
           Surveys*, Vol. 20, No. 3, 1988, pp. 153–189.

[Pnu85]   Pnueli, A.: Linear and Branching Structures in the Semantics and Log-
           ics of Reactive Systems. In: Brauer, W. (ed.): *Automata, Languages
           and Programming (Proc. ICALP'85)*, Nafplion (GR), 1985. LNCS 194,
           Springer-Verlag, Berlin, 1985, pp. 15–32.

[PP92]   Pinna, G. M.; Poigné, A.: On the Nature of Events. Internal report,
           GMD, St. Augustin (D), 1992.

[RBP+91]   Rumbaugh, J.; Blaha, M.; Premerlani, W.; Eddy, F.; Lorensen, W.:
           Object-Oriented Modeling and Design. Prentice-Hall, Englewood Cliffs,
           NJ, 1991.

[RC92]   Rolland, C.; Cauvet, C.: Trends and Perspectives in Conceptual Modeling. In: Loucopoulos, P.; Zicari, R. (eds.): *Conceptual Modeling, Databases, and Case*. John Wiley & Sons, New York, 1992, pp. 27–48.

[Rei85]   Reisig, W.: *Petri Nets*. Springer-Verlag, Berlin, 1985.

[Rei91]   Reisig, W.: Concurrent Temporal Logic. SFB-Bericht 342/7/91 B, TU München, Inst. f. Informatik, 1991.

[RM75]   Roussopoulos, N.; Mylopoulos, J.: Using Semantic Networks for Data Base Management. In: Cardenas, A. (ed.): *Proc. Int. Conf. on Very Large Databases VLDB'75*, Framingham, MA, 1975. pp. 144–157. Reprinted in: Mylopoulos, J.; Brodie, M. (eds.): *Readings in Artificial Intelligence & Databases*, Morgan Kaufmann Publ., San Mateo, 1989.

[RVS91]   Ramackers, G.; Verijn-Stuart, A. A.: Integrating Information System Perspectives with Objects. In: Van Assche, F.; Moulin, B.; Rolland, C. (eds.): *The Object-Oriented Approach in Information Systems*, Québec, Canada, 1991. North-Holland, Amsterdam, 1991, pp. 39–59.

[Saa88]   Saake, G.: *Spezifikation, Semantik und Überwachung von Objektlebensläufen in Datenbanken*. PhD thesis, Technische Universität Braunschweig, 1988. *In German*.

[Saa91]   Saake, G.: Descriptive Specification of Database Object Behaviour. *Data & Knowledge Engineering*, Vol. 6, No. 1, 1991, pp. 47–74.

[Saa93]   Saake, G.: *Objektorientierte Spezifikation von Informationssystemen*. Teubner Verlag, Leipzig, 1993. *In German*.

[Sch92a]   Scholz, S.: Ausnahmebehandlung in Informationssystemen. Diploma Thesis, Techn. Univ. Braunschweig, 1992. *In German*.

[Sch92b]   Schulze, J.: Vereinfachung von dynamischen Objektspezifikationen. Diploma Thesis, Techn. Univ. Braunschweig, 1992. *In German*.

[Sci89]   Sciore, E.: Object Specialization. *ACM Transactions on Information Systems*, Vol. 7, No. 2, 1989, pp. 103–122.

[SE90]   Sernadas, A.; Ehrich, H.-D.: What Is an Object, After All? In: Meersman, R.; Kent, W.; Khosla, S. (eds.): *Object-Oriented Databases: Analysis, Design and Construction (Proc. IFIP WG 2.6 Working Conference DS-4)*, Windermere (UK), 1990. North-Holland, Amsterdam, 1991, pp. 39–70.

[SEC90]   Sernadas, A.; Ehrich, H.-D.; Costa, J.-F.: From Processes to Objects. *The INESC Journal of Research and Development*, Vol. 1, No. 1, 1990, pp. 7–27.

[SF91]   Sernadas, C.; Fiadeiro, J.: Towards Object-Oriented Conceptual Modelling. *Data & Knowledge Engineering*, Vol. 6, 1991, pp. 479–508.

[SFSE88]   Sernadas, A.; Fiadeiro, J.; Sernadas, C.; Ehrich, H.-D.: Abstract Object Types: A Temporal Perspective. In: Banieqbal, B.; Barringer, H.; Pnueli, A. (eds.): *Proc. Colloq. on Temporal Logic in Specification*. LNCS 398, Springer, Berlin, 1988, pp. 324–350.

[SFSE89]   Sernadas, A.; Fiadeiro, J.; Sernadas, C.; Ehrich, H.-D.: The Basic Building Blocks of Information Systems. In: Falkenberg, E.; Lindgreen, P. (eds.): *Information System Concepts: An In-Depth Analysis*, Namur (B), 1989. North-Holland, Amsterdam, 1989, pp. 225–246.

[SJ90]   Saake, G.; Jungclaus, R.: Information about Objects versus Derived Objects. In: Göers, J.; Heuer, A. (eds.): *Second Workshop on Foundations and Languages for Data and Objects*, Aigen (A), 1990. Informatik-Bericht 90/3, Technische Universität Clausthal, pp. 59–70.

[SJ91a]   Saake, G.; Jungclaus, R.: Konzeptioneller Entwurf von Objektgesellschaften. In: Appelrath, H.-J. (ed.): *Proc. Datenbanksysteme in Büro, Technik und Wissenschaft BTW'91*, Kaiserslautern, 1991. Informatik-Fachberichte IFB 270, Springer-Verlag, Berlin, 1991, pp. 327–343. *In German*.

[SJ91b]   Saake, G.; Jungclaus, R.: Specification of Database Dynamics in the TROLL-Language. In: Harper, D.; Norrie, M. (eds.): *Proc. Int. Workshop Specification of Database Systems*, Glasgow, 1991. Springer-Verlag, London, 1992, pp. 228–245.

[SJ92]   Saake, G.; Jungclaus, R.: Views and Formal Implementation in a Three-Level Schema Architecture for Dynamic Objects. In: Gray, P. M. D.; Lucas, R. J. (eds.): *Advanced Database Systems : Proc. 10th British National Conference on Databases (BNCOD 10)*, Aberdeen (Scotland), 1992. Springer, LNCS 618, Berlin, 1992, pp. 78–95.

[SJE91]   Saake, G.; Jungclaus, R.; Ehrich, H.-D.: Object-Oriented Specification and Stepwise Refinement. In: de Meer, J.; Heymer, V.; Roth, R. (eds.): *Proc. Int. IFIP Workshop on Open Distributed Processing*,

IFIP Transactions C: Communication Systems, Vol. 1, Berlin, 1991. North-Holland, Amsterdam, 1992, pp. 99–121.

[SJH93]   Saake, G.; Jungclaus, R.; Hartmann, T.: Application Modelling in Heterogeneous Environments using an Object Specification Language. In: Huhns, M.; Papazoglou, M.P.; Schlageter, G. (eds.): *Int. Conf. on Intelligent & Cooperative Information Systems (ICICIS'93)*, Rotterdam (NL), 1993. IEEE Computer Society Press, 1993, pp. 309–318.

[SM92]   Shlaer, S.; Mellor, S. J.: *Object Lifecycles: Modeling the World in States.* Yourdon Press/Prentice Hall, Englewood Cliffs, NJ, 1992.

[SRGS92]   Sernadas, C.; Resende, P.; Gouveia, P.; Sernadas, A.: In-The-Large Object-Oriented Design of Information Systems. In: Falkenberg, E.; Rolland, C.; El-Sayed, E. N. (eds.): *Information Systems Concepts: Improving the Understanding*, Alexandria, 1992. IFIP Transactions A, North-Holland, Amsterdam, 1992.

[SS77]   Smith, J. M.; Smith, D. C. P.: Database Abstractions: Aggregation and Generalization. *ACM Transactions on Database Systems*, Vol. 2, No. 2, 1977, pp. 105–173.

[SS91]   Saake, G.; Sernadas, A. (eds.): *Information Systems—Correctness and Reusability. Workshop IS-CORE '91, ESPRIT BRA WG 3023, Selected Papers, London, 1991.* TU Braunschweig, Informatik-Bericht 91-03, 1991.

[SS92]   Schwiderski, S.; Saake, G.: Monitoring Temporal Permissions using Partially Evaluated Transition Graphs. In: Lipeck, U. W.; Thalheim, B. (eds.): *Modelling Database Dynamics (Int. Workshop, Selected Papers)*, Volkse (D), 1992. Springer-Verlag, London, 1993, pp. 196–217.

[SS93]   Sernadas, A.; Sernadas, C.: Denotational Semantics of Object Specification within an Arbitrary Temporal Logic Institution. Research report, INESC/DMIST, Lisbon (P), 1993.

[SSC92]   Sernadas, A.; Sernadas, C.; Costa, J. F.: Object Specification Logic. Research report, INESC/DMIST, Lisbon (P), 1992. *To appear in Journal of Logic and Computation.*

[SSE87]   Sernadas, A.; Sernadas, C.; Ehrich, H.-D.: Object-Oriented Specification of Databases: An Algebraic Approach. In: Hammerslay, P. (ed.):

Proc. 13th Int. Conf. on Very Large Databases VLDB'87, Brighton (GB), 1987. Morgan-Kaufmann, Palo Alto, 1987, pp. 107–116.

[SSG+91] Sernadas, A.; Sernadas, C.; Gouveia, P.; Resende, P.; Gouveia, J.: OBLOG—Object-Oriented Logic: An Informal Introduction. Internal report, INESC, Lisbon, 1991.

[Ste92] Stein, A.: Implementierung eines Parsers für eine objektorientierte Spezifikationssprache für Informationssysteme. Diploma Thesis, Techn. Univ. Braunschweig, 1992. *In German.*

[Stö92] Stöcker, R.: Kontextsensitive Analyse von TROLL-Spezifikationen. Diploma Thesis, Techn. Univ. Braunschweig, 1992. *In German.*

[TM87] Turski, W. M.; Maibaum, T. S. E.: *The Specification of Computer Programs.* Addison-Wesley, Reading, MA, 1987.

[TWBL90] Theodoulidis, C.; Wangler, B.; Bubenko, J. A.; Loucopoulos, P.: A Conceptual Model for Temporal Database Applications. SYSLAB Report 71, SYSLAB, Stockholm University, Stockholm, 1990.

[UD86] Urban, S. D.; Delcambre, L.: An Analysis of the Structural, Dynamic, and Temporal Aspects of Semantic Data Models. In: *Proc. Int. Conf. on Data Engineering*, Los Angeles, 1986. ACM, New York, 1986, pp. 382–387.

[Ver91] Verharen, E.: Object-Oriented System Development—An Overview. In: Saake, G.; Sernadas, A. (eds.): *Information Systems—Correctness and Reusability.* Informatik-Bericht 91–03, TU Braunschweig, 1991.

[VL89] Van Horebeek, I.; Lewi, J.: *Algebraic Specifications in Software Engineering.* Springer-Verlag, Berlin, 1989.

[Weg90] Wegner, P.: Concepts and Paradigms of Object-Oriented Programming. *ACM SIGPLAN OOP Messenger*, Vol. 1, No. 1, 1990, pp. 7–87.

[Wie90] Wieringa, R. J.: *Algebraic Foundations for Dynamic Conceptual Models.* PhD thesis, Vrije Universiteit, Amsterdam, 1990.

[Wie91a] Wieringa, R.: Object-Oriented Analysis, Structured Analysis, and Jackson System Development. In: Van Assche, F.; Moulin, B.; Rolland, C. (eds.): *The Object-Oriented Approach in Information Systems*, Québec, Canada, 1991. North-Holland, Amsterdam, 1991.

[Wie91b]   Wieringa, R.: Steps Towards a Method for the Formal Modeling of Dynamic Objects. *Data & Knowledge Engineering*, Vol. 6, 1991.

[Wie91c]   Wieringa, R. J.: A Conceptual Model Specification Language (CMSL Version 2). Technical Report IR–248, Vrije Universiteit, Amsterdam, 1991.

[Win89]   Winskel, G.: An Introduction to Event Structures. In: Bakker, J. de; Roever, W. de; Rozenberg, G. (eds.): *Linear Time, Branching Time and Partial Order in Logics and Models for Concurrency*. LNCS 354, Springer-Verlag, Berlin, 1989, pp. 364–397.

[Wir90]   Wirsing, M.: Algebraic Specification. In: Van Leeuwen, J. (ed.): *Handbook of Theoretical Computer Science, Vol. B: Formal Models and Semantics*. Elsevier Sci. Publ. B.V., Amsterdam, 1990, pp. 675–788.

[WJ91]   Wieringa, R.; Jonge, W. de: The Identification of Objects and Roles— Object Identifiers Revisited. Technical Report IR–267, Vrije Universiteit, Amsterdam, 1991.

[Won81]   Wong, H. K. T.: *Design and Verification of Interactive Information Systems Using TAXIS*. PhD thesis, Computer Systems Research Group, University of Toronto, 1981. *Also as Technical Report CSRG-129*.

[ZM89]   Zdonik, S. B.; Maier, D. (eds.): *Readings in Object-Oriented Database Systems*. Morgan-Kaufmann, Palo Alto, CA, 1989.

Index

Open Distributed Systems

On Concepts, Methods, and Design from a Logical Point of View

by Reinhard Gotzhein

1993. xviii, 227 pages. (Vieweg Advanced Studies in Computer Science) Softcover
ISBN 3-528-05358-5

Presents a new, abstract, and comprehensive view of open distributed systems. Starting point is a small number of core concepts and basic principles, which are informally introduced and precisely defined using mathematical logic. It is shown how the basic concepts of Open Systems Interconnection (OSI) and Open Distributed Processing (ODP), which are currently the most important standardization activities in the context of open distributed systems, can be obtained by specialization and extension of these basic concepts. Application examples include the formal treatment of the interaction point concept and the hierarchical development of communication systems.

This book is a contribution to the field of software engineering in general and to the design of open distributed systems in particular. It is oriented towards the design and implementation of real systems and brings together both formal logical reasoning and current software engineering practice.

Vieweg Publishing · P.O. Box 58 29 · 65048 Wiesbaden